CCD PRECISION PHOTOMETRY WORKSHOP

A SERIES OF BOOKS ON RECENT DEVELOPMENTS IN ASTRONOMY AND ASTROPHYSICS

Managing Editor, D. Harold McNamara
Production Manager, Enid L. Livingston

A.S.P. CONFERENCE SERIES PUBLICATIONS COMMITTEE

Sallie Baliunas, Chair
Carol Ambruster
Catharine Garmany
Mark S. Giampapa
Kenneth Janes

© Copyright 1999 Astronomical Society of the Pacific
390 Ashton Avenue, San Francisco, California 94112

All rights reserved

Printed by Sheridan Books, Inc.

First published 1999

Library of Congress Catalog Card Number: 99-67173
ISBN 1-58381-015-3

Please contact proper address for information on:

PUBLISHING:
Managing Editor
PO Box 24463
211 KMB
Brigham Young University
Provo, UT 84602-4463 USA

Phone: 801-378-2298
Fax: 801-378-4049
E-mail: pasp@astro.byu.edu

ORDERING BOOKS:
Astronomical Society of the Pacific
CONFERENCE SERIES
390 Ashton Avenue
San Francisco, CA 94112 - 1722 USA

Phone: 415-337-2624
Fax: 415-337-5205
E-mail: catalog@aspsky.org
Web Site: http://www.aspsky.org

A SERIES OF BOOKS ON RECENT DEVELOPMENTS IN ASTRONOMY AND ASTROPHYSICS

Vol. 1-Progress and Opportunities in Southern Hemisphere Optical Astronomy: CTIO 25th Anniversary Symposium
ed. V. M. Blanco and M. M. Phillips   ISBN 0-937707-18-X

Vol. 2-Proceedings of a Workshop on Optical Surveys for Quasars
ed. P. S. Osmer, A. C. Porter, R. F. Green, and C. B. Foltz   ISBN 0-937707-19-8

Vol. 3-Fiber Optics in Astronomy
ed. S. C. Barden   ISBN 0-937707-20-1

Vol. 4-The Extragalactic Distance Scale: Proceedings of the ASP 100th Anniversary Symposium
ed. S. van den Bergh and C. J. Pritchet   ISBN 0-937707-21-X

Vol. 5-The Minnesota Lectures on Clusters of Galaxies and Large-Scale Structure
ed. J. M. Dickey   ISBN 0-937707-22-8

Vol. 6-Synthesis Imaging in Radio Astronomy: A Collection of Lectures from the Third NRAO Synthesis Imaging Summer School
ed. R. A. Perley, F. R. Schwab, and A. H. Bridle   ISBN 0-937707-23-6

Vol. 7-Properties of Hot Luminous Stars: Boulder-Munich Workshop
ed. C. D. Garmany   ISBN 0-937707-24-4

Vol. 8-CCDs in Astronomy
ed. G. H. Jacoby   ISBN 0-937707-25-2

Vol. 9-Cool Stars, Stellar Systems, and the Sun. Sixth Cambridge Workshop
ed. G. Wallerstein   ISBN 0-937707-27-9

Vol. 10-Evolution of the Universe of Galaxies: Edwin Hubble Centennial Symposium
ed. R. G. Kron   ISBN 0-937707-28-7

Vol. 11-Confrontation Between Stellar Pulsation and Evolution
ed. C. Cacciari and G. Clementini   ISBN 0-937707-30-9

Vol. 12-The Evolution of the Interstellar Medium
ed. L. Blitz   ISBN 0-937707-31-7

Vol. 13-The Formation and Evolution of Star Clusters
ed. K. Janes   ISBN 0-937707-32-5

Vol. 14-Astrophysics with Infrared Arrays
ed. R. Elston   ISBN 0-937707-33-3

Vol. 15-Large-Scale Structures and Peculiar Motions in the Universe
ed. D. W. Latham and L. A. N. da Costa   ISBN 0-937707-34-1

Vol. 16-Proceedings of the 3rd Haystack Observatory Conference on Atoms, Ions and Molecules: New Results in Spectral Line Astrophysics
ed. A. D. Haschick and P. T. P. Ho   ISBN 0-937707-35-X

Vol. 17-Light Pollution, Radio Interference, and Space Debris
ed. D. L. Crawford   ISBN 0-937707-36-8

Vol. 18-The Interpretation of Modern Synthesis Observations of Spiral Galaxies
ed. N. Duric and P. C. Crane   ISBN 0-937707-37-6

Vol. 19-Radio Interferometry: Theory, Techniques, and Applications, IAU Colloquium 131
ed. T. J. Cornwell and R. A. Perley   ISBN 0-937707-38-4

Vol. 20-Frontiers of Stellar Evolution: 50th Anniversary McDonald Observatory (1939-1989)
ed. D. L. Lambert   ISBN 0-937707-39-2

Vol. 21-The Space Distribution of Quasars
ed. D. Crampton   ISBN 0-937707-40-6

Vol. 22-Nonisotropic and Variable Outflows from Stars
ed. L. Drissen, C. Leitherer, and A. Nota   ISBN 0-937707-41-4

Vol. 23-Astronomical CCD Observing and Reduction Techniques
ed. S. B. Howell   ISBN 0-937707-42-4

Vol. 24-Cosmology and Large-Scale Structure in the Universe
ed. R. R. de Carvalho   ISBN 0-937707-43-0

Vol. 25-Astronomical Data Analysis Software and Systems I
ed. D. M. Worrall, C. Biemesderfer, and J. Barnes   ISBN 0-937707-44-9

Vol. 26-Cool Stars, Stellar Systems, and the Sun, Seventh Cambridge Workshop
ed. M. S. Giampapa and J. A. Bookbinder   ISBN 0-937707-45-7

Vol. 27-The Solar Cycle: Proceedings of the National Solar Observatory/Sacramento Peak 12th Summer Workshop
ed. K. L. Harvey   ISBN 0-937707-46-5

Vol. 28-Automated Telescopes for Photometry and Imaging
ed. S. J. Adelman, R. J. Dukes, Jr., and C. J. Adelman   ISBN 0-937707-47-3

Vol. 29-Viña Del Mar Workshop on Catacysmic Variable Stars
ed. N. Vogt   ISBN 0-937707-48-1

Vol. 30-Variable Stars and Galaxies
ed. B. Warner   ISBN 0-937707-49-X

Vol. 31-Relationships Between Active Galactic Nuclei and Starburst Galaxies
ed. A. V. Filippenko   ISBN 0-937707-50-3

Vol. 32-Complementary Approaches to Double and Multiple Star Research, IAU Colloquium 135
ed. H. A. McAlister and W. I. Hartkopf   ISBN 0-937707-51-1

Vol. 33-Research Amateur Astronomy
ed. S. J. Edberg   ISBN 0-937707-52-X

Vol. 34-Robotic Telescopes in the 1990s
ed. A. V. Filippenko   ISBN 0-937707-53-8

Vol. 35-Massive Stars: Their Lives in the Interstellar Medium
ed. J. P. Cassinelli and E. B. Churchwell   ISBN 0-937707-54-6

Vol. 36-Planets Around Pulsars
ed. J. A. Phillips, S. E. Thorsett, and S. R. Kulkarni   ISBN 0-937707-55-4

Vol. 37-Fiber Optics in Astronomy II
ed. P. M. Gray   ISBN 0-937707-56-2

Vol. 38-New Frontiers in Binary Star Research: Pacific Rim Colloquium
ed. K. C. Leung and I.-S. Nha   ISBN 0-937707-57-0

Vol. 39-The Minnesota Lectures on the Structure and Dynamics of the Milky Way
ed. Roberta M. Humphreys   ISBN 0-937707-58-9

Vol. 40-Inside the Stars, IAU Colloquium 137
ed. Werner W. Weiss and Annie Baglin   ISBN 0-937707-59-7

Vol. 41-Astronomical Infrared Spectroscopy: Future Observational Directions
ed. Sun Kwok   ISBN 0-937707-60-0

Vol. 42-GONG 1992: Seismic Investigation of the Sun and Stars
ed. Timothy M. Brown   ISBN 0-937707-61-9

Vol. 43-Sky Surveys: Protostars to Protogalaxies
ed. B. T. Soifer   ISBN 0-937707-62-7

Vol. 44-Peculiar Versus Normal Phenomena in A-Type and Related Stars, IAU Colloquium 138
ed. M. M. Dworetsky, F. Castelli, and R. Faraggiana   ISBN 0-937707-63-5

Vol. 45-Luminous High-Latitude Stars
ed. D. D. Sasselov   ISBN 0-937707-64-3

Vol. 46-The Magnetic and Velocity Fields of Solar Active Regions, IAU Colloquium 141
ed. H. Zirin, G. Ai, and H. Wang   ISBN 0-937707-65-1

Vol. 47-Third Decennial US-USSR Conference on SETI
ed. G. Seth Shostak   ISBN 0-937707-66-X

Vol. 48-The Globular Cluster-Galaxy Connection
ed. Graeme H. Smith and Jean P. Brodie   ISBN 0-937707-67-8

Vol. 49-Galaxy Evolution: The Milky Way Perspective
ed. Steven R. Majewski   ISBN 0-937707-68-6

Vol. 50-Structure and Dynamics of Globular Clusters
ed. S. G. Djorgovski and G. Meylan   ISBN 0-937707-69-4

Vol. 51-Observational Cosmology
ed. G. Chincarini, A. Iovino, T. Maccacaro, and D. Maccagni   ISBN 0-937707-70-8

Vol. 52-Astronomical Data Analysis Software and Systems II
ed. R. J. Hanisch, R. J. V. Brissenden, and Jeannette Barnes   ISBN 0-937707-71-6

Vol. 53-Blue Stragglers
ed. Rex A. Saffer   ISBN 0-937707-72-4

Vol. 54-The First Stromlo Symposium: The Physics of Active Galaxies
ed. Geoffrey V. Bicknell, Michael A. Dopita, and Peter J. Quinn   ISBN 0-937707-73-2

Vol. 55-Optical Astronomy from the Earth and Moon
ed. Diane M. Pyper and Ronald J. Angione   ISBN 0-937707-74-0

Vol. 56-Interacting Binary Stars
ed. Allen W. Shafter   ISBN 0-937707-75-9

Vol. 57-Stellar and Circumstellar Astrophysics
ed. George Wallerstein and Alberto Noriega-Crespo   ISBN 0-937707-76-7

Vol. 58-The First Symposium on the Infrared Cirrus and Diffuse Interstellar Clouds
ed. Roc M. Cutri and William B. Latter   ISBN 0-937707-77-5

Vol. 59-Astronomy with Millimeter and Submillimeter Wave Interferometry, IAU Colloquium 140
ed. M. Ishiguro and Wm. J. Welch   ISBN 0-937707-78-3

Vol. 60-The MK Process at 50 Years: A Powerful Tool for Astrophysical Insight: A Workshop of the Vatican Observatory
ed. C. J. Corbally, R. O. Gray, and R. F. Garrison   ISBN 0-937707-79-1

Vol. 61-Astronomical Data Analysis Software and Systems III
ed. Dennis R. Crabtree, R. J. Hanisch, and Jeannette Barnes   ISBN 0-937707-80-5

Vol. 62-The Nature and Evolutionary Status of Herbig Ae / Be Stars
ed. P. S. Thé, M. R. Pérez, and E. P. J. van den Heuvel   ISBN 0-937707-81-3

Vol. 63-Seventy-Five Years of Hirayama Asteroid Families: The Role of Collisions in the Solar System History
ed. Y. Kozai, R. P. Binzel, and T. Hirayama   ISBN 0-937707-82-1

Vol. 64-Cool Stars, Stellar Systems, and the Sun, Eighth Cambridge Workshop
ed. Jean-Pierre Caillault   ISBN 0-937707-83-X

Vol. 65-Clouds, Cores, and Low Mass Stars
ed. Dan P. Clemens and Richard Barvainis ISBN 0-937707-84-8

Vol. 66- Physics of the Gaseous and Stellar Disks of the Galaxy
ed. Ivan R. King ISBN 0-937707-85-6

Vol. 67-Unveiling Large-Scale Structures Behind the Milky Way
ed. C. Balkowski and R. C. Kraan-Korteweg ISBN 0-937707-86-4

Vol. 68-Solar Active Region Evolution: Comparing Models with Observations
ed. K. S. Balasubramaniam and George W. Simon ISBN 0-937707-87-2

Vol. 69-Reverberation Mapping of the Broad-Line Region in Active Galactic Nuclei
ed. P. M. Gondhalekar, K. Horne, and B. M. Peterson ISBN 0-937707-88-0

Vol. 70-Groups of Galaxies
ed. Otto G. Richter and Kirk Borne ISBN 0-937707-89-9

Vol. 71-Tridimensional Optical Spectroscopic Methods in Astrophysics, IAU Colloquium 149
ed. G. Comte and M. Marcelin ISBN 0-937707-90-2

Vol. 72-Millisecond Pulsars: A Decade of Surprise.
ed. A. A. Fruchter, M. Tavani, and D. C. Backer ISBN 0-937707-91-0

Vol. 73-Airborne Astronomy Symposium on the Galactic Ecosystem: From Gas to Stars to Dust
ed. M. R. Haas, J. A. Davidson, and E. F. Erickson ISBN 0-937707-92-9

Vol. 74-Progress in the Search for Extraterrestrial Life: 1993 Bioastronomy Symposium
ed. G. Seth Shostak ISBN 0-937707-93-7

Vol. 75-Multi-Feed Systems for Radio Telescopes
ed. D. T. Emerson and J. M. Payne ISBN 0-937707-94-5

Vol. 76-GONG '94: Helio- and Astero-Seismology from the Earth and Space
ed. Roger K. Ulrich, Edward J. Rhodes, Jr., and Werner Däppen ISBN 0-937707-95-3

Vol. 77-Astronomical Data Analysis Software and System IV
ed. R. A. Shaw, H. E. Payne, and J. J. E. Hayes ISBN 0-937707-96-1

Vol. 78-Astrophysical Applications of Powerful New Databases: Joint Discussion No. 16 of the 22nd General Assembly of the IAU
ed. S. J. Adelman and W. L. Wiese ISBN 0-937707-97-X

Vol. 79-Robotic Telescopes: Current Capabilities, Present Developments, and Future Prospects for Automated Astronomy
ed. Gregory W. Henry and Joel A. Eaton ISBN 0-937707-98-8

Vol. 80-The Physics of the Interstellar Medium and Intergalactic Medium
ed. A. Ferrara, C. F. McKee, C. Heiles, and P. R. Shapiro ISBN 0-937707-99-6

Vol. 81-Laboratory and Astronomical High Resolution Spectra
ed. A. J. Sauval, R. Blomme, and N. Grevesse ISBN 1-886733-01-5

Vol. 82-Very Long Baseline Interferometry and the VLBA
ed. J. A. Zensus, P. J. Diamond, and P. J. Napier ISBN 1-886733-02-3

Vol. 83-Astrophysical Applications of Stellar Pulsation. IAU Colloquium 155
ed. R. S. Stobie and P. A. Whitelock ISBN 1-886733-03-1

Vol. 84-The Future Utilisation of Schmidt Telescopes, IAU Colloquium 148
ed. Jessica Chapman, Russell Cannon, Sandra Harrison, and Bambang Hidayat ISBN 1-886733-05-8

Vol. 85-Cape Workshop on Magnetic Cataclysmic Variables
ed. D. A. H. Buckley and B. Warner ISBN 1-886733-06-6

Vol. 86-Fresh Views of Elliptical Galaxies
ed. Alberto Buzzoni, Alvio Renzini, and Alfonso Serrano ISBN 1-886733-07-4

Vol. 87-New Observing Modes for the Next Century
ed. Todd Boroson, John Davies, and Ian Robson ISBN 1-886733-08-2

Vol. 88- Clusters, Lensing, and the Future of the Universe
ed. Virginia Trimble and Andreas Reisenegger ISBN 1-886733-09-0

Vol. 89-Astronomy Education: Current Developments, Future Coordination
ed. John R. Percy ISBN 1-886733-10-4

Vol. 90-The Origins, Evolution, and Destinies of Binary Stars in Clusters
ed. E. F. Milone and J. C. Mermilliod ISBN 1-886733-11-2

Vol. 91-Barred Galaxies, IAU Colloquium 157
ed. R. Buta, D. A. Crocker, and B. G. Elmegreen ISBN 1-886733-12-0

Vol. 92-Formation of the Galactic Halo–Inside and Out
ed. H. L. Morrison and A. Sarajedini ISBN 1-886733-13-9

Vol. 93-Radio Emission from the Stars and the Sun
ed. A. R. Taylor and J. M. Paredes ISBN 1-886733-14-7

Vol. 94-Mapping, Measuring, and Modelling the Universe
ed. Peter Coles, Vincent Martinez, and Maria-Jesus Pons-Borderia ISBN 1-886733-15-5

Vol. 95-Solar Drivers of Interplanetary and Terrestrial Disturbances: Proceedings of 16th International Workshop, National Solar Observatory/Sacramento Peak
ed. K. S. Balasubramaniam, S. L. Keil, and R. N. Smartt ISBN 1-886733-16-3

Vol. 96- Hydrogen-Deficient Stars
ed. C. S. Jeffery and U. Heber ISBN 1-886733-17-1

Vol. 97-Polarimetry of the Interstellar Medium
ed. W. G. Roberge and D. C. B. Whittet ISBN 1-886733-18-X

Vol. 98-From Stars to Galaxies: The Impact of Stellar Physics on Galaxy Evolution
ed. Claus Leitherer, Uta Fritze-von Alvensleben, and John Huchra ISBN 1-886733-19-8

Vol. 99-Cosmic Abundances: Proceedings of the 6th Annual October Astrophysics Conference
ed. Stephen S. Holt and Geroge Sonneborn ISBN 1-886733-20-1

Vol. 100-Energy Transport in Radio Galaxies and Quasars
ed. P. E. Hardee, A. H. Bridle, and J. A. Zensus ISBN 1-886733-21-X

Vol. 101-Astronomical Data Analysis Software and Systems V
ed. George H. Jacoby and Jeannette Barnes ISSN 1080-7926

Vol. 102-The Galactic Center, 4th ESO/CTIO Workshop
ed. Roland Gredel ISBN 1-886733-22-8

Vol. 103-The Physics of Liners in View of Recent Observations
ed. M. Eracleous, A. Koratkar, C. Leitherer, and L. Ho ISBN 1-886733-23-6

Vol. 104-Physics, Chemistry, and Dynamics of Interplanetary Dust, IAU Colloquium 150
ed. Bo A. S. Gustafson and Martha S. Hanner ISBN 1-886733-24-4

Vol. 105-Pulsars: Problems and Progress, IAU Colloquium 160
ed. M. Bailes, S. Johnston, and M. A. Walker ISBN 1-886733-25-2

Vol. 106-Minnesota Lectures on Extragalactic Neutral Hydrogen
ed. Evan D. Skillman ISBN 1-886733-26-0

Vol. 107-Completing the Inventory of the Solar System: A Symposium held in conjuunction with the 106th Annual Meeting of the ASP
ed. Terrence W. Rettig and Joseph M. Hahn ISBN 1-886733-27-9

Vol. 108-M. A. S. S. Model Atmospheres and Spectrum Synthesis: 5th Vienna Workshop
ed. S. J. Adelman, F. Kupka, and W. W. Weiss ISBN 1-886733-28-7

Vol. 109-Cool Stars, Stellar Systems, and the Sun, Ninth Cambridge Workshop
ed. Roberto Pallavicini and Andrea K. Dupree ISBN 1-886733-29-5

Vol. 110-Blazar Continuum Variability
ed. H. R. Miller, J. R. Webb, and J. C. Noble ISBN 1-886733-30-9

Vol. 111-Magnetic Reconnection in the Solar Atmosphere: Proceedings of a Yohkoh Conference
ed. R. D. Bentley and J. T. Mariska ISBN 1-886733-31-7

Vol. 112-The History of the Milky Way and Its Satellite System
ed. A. Burkert, D. H. Hartmann, and S. R. Majewski ISBN 1-886733-32-5

Vol. 113-Emission Lines in Active Galaxies: New Methods and Techniques, IAU Colloquium 159
ed. B. M. Peterson, F. Z. Cheng, and A. S. Wilson ISBN 1-886733-33-3

Vol. 114-Young Galaxies and QSO Absorption-Line Systems
ed. Sueli M. Viegas, Ruth Gruenwald, and Reinaldo R. de Carvalho ISBN 1-886733-34-1

Vol. 115-Galactic and Cluster Cooling Flows
ed. Noam Soker ISBN 1-886733-35-X

Vol. 116-The Second Stromlo Symposium: The Nature of Elliptical Galaxies
ed. M. Arnaboldi, G. S. Da Costa, and P. Saha ISBN 1-886733-36-8

Vol. 117- Dark and Visible Matter in Galaxies
ed. Massimo Persic and Paolo Salucci ISBN 1-886733-37-6

Vol. 118-First Advances in Solar Physics Euroconference: Advances in the Physics of Sunspots
ed. B. Schmieder, J. C. del Toro Iniesta, and M. Vázquez ISBN 1-886733-38-4

Vol. 119-Planets Beyond the Solar System and the Next Generation of Space Missions
ed. David R. Soderblom ISBN 1-886733-39-2

Vol. 120-Luminous Blue Variables: Massive Stars in Transition
ed. Antonella Nota and Henny J. G. L. M. Lamers ISBN 1-886733-40-6

Vol. 121-Accretion Phenomena and Related Outflows, IAU Colloquium 163
ed. D. T. Wickramasinghe, G. V. Bicknell and L. Ferrario ISBN 1-886733-41-4

Vol. 122-From Stardust to Planetesimals: Symposium held as part of the 108th Annual Meeting of the ASP
ed. Yvonne J. Pendleton and A. G. G. M. Tielens ISBN 1-886733-42-2

Vol. 123-The 12th 'Kingston Meeting': Computational Astrophysics
ed. David A. Clarke and Michael J. West ISBN 1-886733-43-0

Vol. 124-Diffuse Infrared Radiation and the IRTS
ed. Haruyuki Okuda, Toshio Matsumoto, and Thomas L. Roellig ISBN 1-886733-44-9

Vol. 125- Astronomical Data Analysis Software and Systems VI
ed. Gareth Hunt and H. E. Payne ISBN 1-886733-45-7

Vol. 126-From Quantum Fluctuations to Cosmological Structures
ed. D. Vallis-Gabaud, M. A. Hendry, P. Molaro, and K. Chamcham ISBN 1-886733-46-5

Vol. 127-Proper Motions and Galactic Astronomy
ed. Roberta M. Humphreys ISBN 1-886733-47-3

Vol. 128- Mass Ejection from AGN (Active Galactic Nuclei)
ed. N. Arav, I. Shlosman, and R. J. Weymann   ISBN 1-886733-48-1

Vol. 129-The George Gamow Symposium
ed. E. Harper, W. C. Parke, and G. D. Anderson   ISBN 1-886733-49-X

Vol. 130-The Third Pacfic Rim Conference on Recent Development on Binary Star Research
ed. Kam-Ching Leung   ISBN 1-886733-50-3

Vol. 131-Boulder-Munich II: Properties of Hot, Luminous Stars
ed. Ian D. Howarth   ISBN 1-886733-51-1

Vol. 132-Star Formation with the Infrared Space Observatory (ISO)
ed. João L. Yun and René Liseau   ISBN 1-886733-53-X

Vol. 133-Science with the NGST
ed. Eric P. Smith and Anuradha Koratkar   ISBN 1-886733-53-8

Vol. 134-Brown Dwarfs and Extrasolar Planets
ed. Rafael Rebolo, Eduardo L. Martin,
and Maria Rosa Zapatero Osorio   ISBN 1-886733-54-6

Vol. 135-A Half Century of Stellar Pulsation Interpretations: A Tribute to Arthur N. Cox
ed. P. A Bradley and J. A. Guzik   ISBN 1-886733-55-4

Vol. 136- Galactic Halos: A UC Santa Cruz Workshop
ed. Dennis Zaritsky   ISBN 1-886733-56-2

Vol. 137-Wild Stars in the Old West: Proceedings of the 13th North American Workshop
on Cataclysmic Variables and Related Objects
ed. S. Howell, E.Kuulkers, and C. Woodward   ISBN 1-886733-57-0

Vol. 138-1997 Pacific Rim Conference on Stellar Astrophysics
ed. Kwing L. Chan, K. S. Cheng, and Harinder P. Singh   ISBN 1-886733-58-9

Vol. 139-Preserving the Astronomical Windows, Proceedings of Joint Discussion
No. 5 of the 23rd General Assembly of the IAU
ed. Syuzo Isobe and Tomohiro Hirayama   ISBN 1-886733-59-7

Vol. 140-Synoptic Solar Physics – 18th NSO/Sacramento Peak Summer Workshop
ed. K. S. Balasubramaniam, J. W. Harvey, and D. M. Rabin   ISBN 1-886733-60-0

Vol. 141-Astrophysics From Antarctica
A Symposium held as a part of the 109th Annual Meeting of the ASP
ed. Giles Novak and Randall H. Landsberg   ISBN 1-886733-61-9

Vol. 142-The Stellar Initial Mass Function, 38th Herstmonceux Conference
ed. Gerry Gilmore and Debbie Howell   ISBN 1-886733-62-7

Vol. 143-The Scientific Impact of the Goddard High Resolution Spectrograph
ed. John C. Brandt, Thomas B. Ake III, and Carolyn Collins Petersen   ISBN 1-886733-63-5

Vol. 144- Radio Emission from Galactic and Extragalactic Compact Sources,
IAU Colloquium 164
ed. J. Anton Zensus, G. B. Taylor, and J. M. Wrobel   ISBN 1-886733-64-3

Vol. 145-Astronomical Data Analysis Software and Systems VII
ed. Rudolf Albrecht, Richard N. Hook, and Howard A. Bushouse   ISBN 1-886733-65-1

Vol. 146-The Young Universe: Galaxy Formation and Evolution at
Intermediate and High Redshift
ed. S. D'Odorico, A. Fontana, and E. Giallongo   ISBN 1-886733-66-X

Vol. 147-Abundance Profiles: Diagnostic Tools for Galaxy History
ed. Daniel Friedli, Mike Edmunds, Carmelle Robert,
and Laurent Drissen   ISBN 1-886733-67-8

Vol. 148-Origins
ed. Charles E. Woodward, J. Michael Shull,
and Harley A. Thronson, Jr. ISBN 1-886733-68-6

Vol. 149-Solar System Formation and Evolution
ed. D. Lazzaro, R. Vieira Martins, S. Ferraz-Mello,
J. Fernández, and C. Beaugé ISBN 1-886733-69-4

Vol. 150-New Perspectives on Solar Prominences, IAU Colloquium 167
ed. David Webb, David Rust, and Brigitte Schmieder ISBN 1-886733-70-8

Vol. 151-Cosmic Microwave Background and Large Scale Structure of the Universe
ed. Yong-Ik Byun and Kin-Wang Ng ISBN 1-886733-71-6

Vol. 152-Fiber Optics in Astronomy III
ed. S. Arribas, E. Mediavilla, and F. Watson ISBN 1-886733-72-4

Vol. 153-Library and Information Services in Astronomy III, (LISA III)
ed. Uta Grothkopf, Heinz Andernach, Sarah Stevens-Rayburn,
and Monique Gomez ISBN 1-886733-73-2

Vol. 154-Cool Stars, Stellar Systems, and the Sun, Tenth Cambridge Workshop
ed. Robert A. Donahue and Jay A. Bookbinder ISBN 1-886733-74-0

Vol. 155-Second Advances in Solar Physics Euroconference:
Three-Dimensional Structure of Solar Active Regions
ed. Costas E. Alissandrakis and Brigitte Schmieder ISBN 1-886733-75-9

Vol. 156-Highly Redshifted Radio Lines
ed. C. L. Carilli, S. J. E. Radford, K. M. Menten and
G. I. Langston ISBN 1-886733-76-7

Vol. 157-Annapolis Workshop on Magnetic Cataclysmic Variables
ed. Coel Hellier and Koji Mukai ISBN 1-886733-77-5

Vol. 158-Solar and Stellar Activity: Similarities and Differences
ed. C. J. Butler and J. G. Doyle ISBN 1-886733-78-3

Vol. 159-BL Lac Phenomenon
ed. Leo O. Takalo and Aimo Sillanpää ISBN 1-886733-79-1

Vol. 160-Astrophysical Discs, An EC Summer School
ed. J. A. Sellwood and Jeremy Goodman ISBN 1-886733-80-5

Vol. 161-High Energy Processes in Accreting Black Holes
ed. Juri Poutanen and Roland Svensson ISBN 1-886733-81-3

Vol. 162-Quasars and Cosmology
ed. Gary Ferland and Jack Baldwin ISBN 1-886733-83-X

Vol. 163-Star Formation in Early Type Galaxies
ed. Jordi Cepa and Patricia Carral ISBN 1-886733-84-8

Vol. 164-Ultraviolet–Optical Space Astronomy Beyond HST
ed. Jon A. Morse, J. Michael Shull and Anne L. Kinney ISBN 1-886733-85-6

Vol. 165-The Third Stromlo Symposium: The Galactic Halo
ed. Brad K. Gibson, Tim S. Axelrod, and Mary E. Putman ISBN 1-886733-86-4

Vol. 166-Stromlo Workshop on High-Velocity Clouds
ed. Brad K. Gibson and Mary E. Putman ISBN 1-886733-87-2

Vol. 167-Harmonizing Cosmic Distance Scales in a Post-HIPPARCOS Era
ed. Daniel Egret and André Heck ISBN 1-886733-88-0

Vol. 168-New Perspectives on the Interstellar Medium
ed. A. R. Taylor, T. L. Landecker, and G. Joncas ISBN 1-886733-89-9

Vol. 169-11th European Workshop on White Dwarfs
ed. J.-E. Solheim and E. G. Meistas ISBN 1-886733-91-0

Vol. 170-The Low Surface Brightness Universe, IAU Colloquium 171
ed. J. I. Davies, C. Impey, and S. Phillipps ISBN 1-886733-92-9

Vol. 171-LiBeB, Cosmic Rays, and Related X-and Gamma-Rays
ed. Reuven Ramaty, Elisabeth Vangioni-Flam,
Michel Cassé, and Keith Olive ISBN 1-886733-93-7

Vol. 172-Astronomical Data Analysis Software and Systems VIII
ed. David M. Mehringer, Raymond L. Plante, and
Douglas A. Roberts ISBN 1-886733-94-5

Vol. 173-Theory and Tests of Convection In Stellar Structure
ed. Álvaro Giménez, Edward F. Guinan, and Benjamín Montesinos ISBN 1-886733-95-3

Vol. 174-Catching the Perfect Wave: Adaptive Optics and Interferometry for the 21st
Century, A Symposium held as a part of the 110th Annual Meeting of the ASP
ed. Sergio R. Restaino, William Junor, and Nebojsa Duric ISBN 1-886733-96-1

Vol. 175-Structure and Kinematics of Quasar Broad Line Regions
ed. C. M. Gaskell, W. N. Brandt, M. Dietrich, D. Dultzin-Hacyan,
and M. Eracleous ISBN 1-886733-97-X

Vol. 176-Observational Cosmology: The Development of Galaxy Systems
ed. Giuliano Giuricin, Marino Mezzetti, and Paolo Salucci ISBN 1-58381-000-5

Vol. 177-Astrophysics with Infrared Surveys: A Prelude to SIRTF
ed. Michael D. Bicay, Chas A. Beichman, Roc M. Cutri,
and Barry F. Madore ISBN 1-58381-001-3

Vol. 178-Stellar Dynamos: Nonlinearity and Chaotic Flows
ed. Manuel Núñez and Antonio Ferriz-Mas ISBN 1-58381-002-1

Vol. 179-Eta Carinae At The Millennium
ed. Jon A. Morse, Roberta M. Humphreys, and Augusto Damineli ISBN 1-58381-003-X

Vol. 180-Synthesis Imaging in Radio Astronomy II
ed. G. B. Taylor, C. L. Carilli, and R. A. Perley ISBN 1-58381-005-6

Vol. 181-Microwave Foregrounds
ed. Angelica de Oliveira-Costa and Max Tegmark ISBN 1-58381-006-4

Vol. 182-Galaxy Dynamics, A Rutgers Symposium
ed. David Merritt, J. A. Sellwood, and Monica Valluri ISBN 1-58381-007-2

Vol. 183-High Resolution Solar Physics: Theory, Observations and Techniques
ed. T. R. Rimmele, K. S. Balasubramaniam, and R. R. Radick ISBN 1-58381-009-9

Vol. 184-Third Advances in Solar Physics Euroconference: Magnetic Fields and Oscillations
ed. B. Schmieder, A. Hofmann, and J. Staude ISBN 1-58381-010-2

Vol. 185-Precise Stellar Radial Velocities, IAU Colloquium 170
ed. J. B. Hearnshaw and C. D. Scarfe ISBN 1-58381-011-0

Vol. 186-The Central Parsecs of the Galaxy
ed. Heino Falcke, Angela Cotera, Wolfgang J. Duschl, Fulvio Melia,
and Marcia J. Rieke ISBN 1-58381-012-9

Vol. 187-The Evolution of Galaxies on Cosmological Timescales
ed. J. E. Beckman and T. J. Mahoney — ISBN 1-58381-013-7

Vol. 188-Optical and Infrared Spectroscopy of Circumstellar Matter
ed. Eike W. Guenther, Bringfried Stecklum, and Sylvio Klose — ISBN 1-58381-014-5

Vol. 189-CCD Precision Photometry Workshop
ed. Eric R. Craine, Roy A. Tucker, and Jeannette Barnes — ISBN 1-58381-015-3

Book orders or inquiries concerning these volumes should be directed to the:

Astronomical Society of the Pacific Conference Series
390 Ashton Avenue
San Francisco, CA 94112-1722 USA

Phone: 415-337-2126   E-mail: catalog@aspsky.org
Fax: 415-337-5205   Web Site: http://www.aspsky.org

ASTRONOMICAL SOCIETY OF THE PACIFIC
CONFERENCE SERIES

Volume 189

CCD PRECISION PHOTOMETRY WORKSHOP

Proceedings of a meeting held at
San Diego, California, USA
6-7 June 1998

Edited by

Eric R. Craine
Western Research Company, Inc., Tucson, Arizona, USA
GNAT, Inc., Tucson, Arizona, USA
Department of Physics, Colorado State University, Ft. Collins, Colorado, USA

Roy A. Tucker
National Solar Observatory, Tucson, Arizona, USA

and

Jeannette Barnes
National Optical Astronomy Observatories, Tucson, Arizona, USA

Table of Contents

Preface	xvii
Conference participants	xix
Photometry Issues	3
D. L. Crawford	
Another Look at Open Cluster Photometry: Precision and Accuracy	6
D. L. Crawford	
Photometric Systems	9
D. L. Crawford	
Atmospheric Constraints on Precision Photometry	16
R. J. Angione	
Some Practical Aspects of CCD Camera Construction	24
R. A. Tucker	
Stellar Photometry Tools in IRAF	35
L. E. Davis	
CCD Aperture Photometry	50
K. J. Mighell	
Point-Spread Function Fitting Photometry	56
J. N. Heasley	
Increasing Precision and Accuracy in Photometric Measurements	74
M. V. Newberry	
The GNAT Automatic Imaging Telescope for CCD Photometry	83
E. R. Craine, D. L. Crawford and P. R. Craine	
Observing Blazars with the WEB Telescope	95
J. R. Mattox	
CCD Photometry with the A. R. Cross Telescope of the Rothney Astrophysical Observatory	103
E. F. Milone and P. Langill	
A High Performance CCD System for Detection of Optical Supernova Remnants	111
T. Foster, D. Hube, J. Couch, B. Martin, D. Routledge and F. Vaneldik	
Photometry at the Robotic Lunar Observatory in Flagstaff	125
J. M. Anderson	
GNAT Engineering Issues: Thermal Stability of a Prototype Imaging Photometer	133
P. R. Craine	

A 1.0 meter Automatic Imaging Telescope for CCD Photometry 143
J. D. Wray

ARLO: Automated CCD Reductions for the Construction of an All-sky Catalog of Photometric Calibrators 154
R. Casalegno, B. Bucciarelli, J. Garcia and B. M. Lasker

Photometric Search for Extra-Solar Planets 170
S. B. Howell, M. E. Everett, G. Esquerdo, D. R. Davis, S. Weidenschilling and T. Van Lew

CCDPHOT Photometry of Extremely Metal-Poor Stars 192
B. J. Anthony-Twarog, T. C. Beers, S. L. Hawley, A. Sarajedini and B. A. Twarog

Precise Photometry of the GSC 3493_742 Field 198
R. M. Robb and R. Greimel

Toward Precision Photometry of Red Variable Stars 207
T. A. Ostrowski and R. E. Stencel

Comparative Photometric Reduction Techniques 213
E. R. Craine, M. Snowden and P. Martinez

Differential Photometry Using the GNAT 0.5 m Prototype 238
J. M. Taylor, E. R. Craine and M. S. Giampapa

CCD Strømgren and H_β Photometry of Open Clusters 252
N. Kaltcheva, M. I. Andersen, H. Jønch-Sørensen and A. N. Sørensen

Photometric Characteristics of the Etelman Observatory in St. Thomas, US Virgin Islands 257
D. A. Aurin, J. E. Neff and D. M. Drost

Update on Pine Mountain Observatory (PMO), A Report to GNAT - Summer 1998 260
R. Kang

National Astronomical Observatory Tonantzintla, México 264
J. H. Peña, R. Peniche, B. Sánchez, C. Tejada and R. Costero

Author index 271

Subject index 273

Preface

The Global Network of Astronomical Telescopes (GNAT) was a concept for a longitudinally distributed network of small automated astronomical telescopes which would serve primarily as imaging photometers. The intent was that such a network would be extremely efficient at collecting data, and could be dedicated to specific long-term, large scale, or high observational frequency programs of types which cannot be granted time on more conventional observing facilities. Among the science programs envisioned for GNAT are intensive observations of standard stars, conduct of transit event searches in an effort to find extrasolar planetary systems, and numerous other long-term photometric monitoring programs.

GNAT has placed its first test telescope and is, at this time, actively involved in developing software for both system control and communications and for subsequent data handling and reduction. One of the key issues which emerged early in the program was related to the types of science we could hope to do, based in part on the photometric precision which could be attained with the systems envisioned. This was all to be CCD photometry, and a little research soon showed considerable diversity of results reported by others in the literature, and indeed some astronomers who continue to argue that reasonable CCD photometry cannot be done at all. Countering this extreme were others who make fairly extravagant claims for their work with CCD images.

In order to get a better grasp of what we might hope to attain, and the care which must be invested to obtain those results, it was suggested that GNAT sponsor a CCD precision photometry workshop, which came to fruition in June 1998 with the support and co-sponsorship of the Department of Astronomy at San Diego State University.

It is unfortunate that not all of the activity of that workshop can be conveyed in this volume. In addition to invited and submitted papers, both oral and poster presentation format, this meeting benefited from some extended and very lively discussion sessions, all of which served to demonstrate that there remains considerable disagreement over the best techniques available to attain high precision CCD photometry. Further, for a meeting this small, there was a large number of demonstrations of both hardware and software relevant to the production of high quality CCD photometry.

Roy Tucker and Bob Leach provided informative demonstrations of various hardware systems at the component levels, and discussed effects of hardware configurations on eventual photometric precision. Jeannette Barnes and Lindsey Davis gave much appreciated demonstrations of IRAF and Ken Mighell showed his CCD aperture photometry program. Jim Heasley demonstrated creation of nearly instantaneous cluster color-magnitude diagrams using his SPS CCD photometry program and Mike Newberry showed MIRA. These sessions were very productive and allowed many of the workshop attendees to gain some real hands-on experience with systems which were, in some cases, new to them, as well as to ask questions of the people most intimately connected with the development of these tools.

The formal paper presentations are assembled here, rather lightly edited to preserve their original flavor. These presentations cover a broad range of issues and applications of CCD photometry, and recount many of the problems (and solutions) which have been encountered. We thank all of the authors who made the effort not only to present their ideas to the assembled group, but to convert those ideas to written format for others to share.

We would like to acknowledge David Crawford, of GNAT, for his help and encouragement in conducting the workshop, and for his contributions to it. Ron Angione took on the formidable task of coordinating the meeting for San Diego State University, even though he was simultaneously embroiled in a similar (albeit much larger) effort for the summer American Astronomical Society meeting. Robert Leach generously hosted multiple guided tours of the CCD Imaging Laboratory at San Diego State University. The city of San Diego cooperated with its normal perfect weather, and several impromptu photometry discussions were pleasantly conducted at Anthony's Seafood Grotto on the Embarcadero.

It is our hope that the next GNAT sponsored meeting will be to report on results of GNAT photometry programs. These include work to develop a set of well observed Stromvil system standard stars, preliminary work on GNAT monitoring of potential extra-solar system planets, and monitoring of Mira variables, among others. We expect that all of these programs will be beneficiaries of discussions which led to this current volume.

The Editors
Tucson, Arizona

August, 1999

Cover illustration: The differential R band light curve of GSC 3493_1097 obtained with a 0.5-m automated telescope and CCD imaging photometer (from Robb and Greimel page 198).

Participant List

James M. Anderson, NAU/USGS
Ron Angione, San Diego State University
Barbara Anthony-Twarog, University of Kansas
Dirk Aurin, College of Charlestown
Jeannette Barnes, NOAO
Sydney Barnes, Lowell Observatory
John W. Briggs, Yerkes Observatory
Roberto Casalegno, Osservatorio Astronomico Torino, Italy
Alan Chamberlin, JPL
Patricia Cheng, Cal State Fullerton
Rafael Costero, Institute of Astronomy, Mexico
Eric R. Craine, GNAT
Patrick R. Craine, UC Berkeley
David L. Crawford, GNAT
Roger B. Culver, Colorado State University
Lindsey Davis, NOAO
Don Epand, TechRep, Arizona
Mark Everett, University of Wyoming
Tyler J. Foster, University of Alberta, Canada
John Harbold, San Diego State University
Jim Heasley, University of Hawaii
Arne A. Hendon, USARA/USNO
Steve Howell, University of Wyoming
Nadejda T. Kaltcheva, St. Andrews University, UK
Rick Kang, Pine Mountain Observatory
Robert Leach, San Diego State University
G. W. Lockwood, Lowell Observatory
John R. Mattox, Boston University
Russet McMillan, Apache Point Observatory
Ken Mighell, NOAO
E. F. Milone, University of Calgary, Canada
James Neff, College of Charleston
Burt Nelson, San Diego State University
Mike Newberry, Axiom Research
Therese Ostrowski, University of Denver
Jose H. Peña, University of Mexico

Rosario Peniche, University of Mexico
Hong Rhie Sun, Notre Dame
Russell Robb, University of Victoria
Wayne Rosing, Las Cumbres Observatory
Edward Schmidt, University of Nebraska
William J. Schuster, Institute of Astronomy, Mexico
Micheal Snowden, Lanka Astronomical Observatory
Jeff Stoner, GNAT
F. D. Talbert, San Diego State University
Jacob Taylor, Harvard University
Mark Trueblood, NOAO
Roy A. Tucker, GONG, NOAO
Patricia Van Lew, University of Wyoming
Lawrence H. Wasserman, Lowell Observatory
Paul Weissman, Jet Propulsion Lab
Jim Wray, SciTech Astronomical Research

Photometry Issues

David L. Crawford

GNAT, Inc., 2127 E Speedway, Tucson AZ 85719, crawford@gnat.org

Abstract. There are many other issues than the CCD detector that can lead to errors in photometry. Photometrists must be aware of all of these and must take appropriate actions to minimize such adverse effects. This paper lists some of them.

1. Introduction

Precision and accuracy are key issues in photometry, no matter how the photometry is done. I will discuss some of the photometric issues in this short paper, and a few of the specific issues relative to photometric systems in a second paper. A poster paper gives a few examples of what can happen when the process is not well handled.

2. Doing Photometry

It is possible to consider photometry as the application of the following equation:

$$I(\lambda,t) = c \int (I_s - I_b)\, IM\ A\ T\ P\ F\ D\ O\, d\lambda \tag{1}$$

where $I(\lambda,t)$ is the flux we observe and c is a normalizing constant. The other terms are discussed one at a time below.

I_s is the item that we are usually trying to determine, the flux from the star (or other object) that we are observing. However, sometimes we are just using a star as a probe to determine something else, such as atmospheric extinction or interstellar reddening values in the direction of observation.

Most all of the terms in the above equation are wavelength dependent, vary with time, location in the field of view, and with other factors such as temperature, humidity, telescope position, site location and altitude, and so forth. Identifying, calibrating, and otherwise correcting for all these multitude of adverse effects are critical. The pitfalls for systematic and accidental errors are huge. Errors in any one term can really cause problems. Both precision and accuracy require that these effects all be understood, calibrated, and corrected adequately. That takes time and effort, and often such time is not available. Without doing the job right, we are lucky to get the "right" answer.

I_s (λ, t, x, y, ...) is the flux emitted by the star or other object we are observing. This can and does include the continuous and line emission and absorption lines, bands, breaks, the effects of any circumstellar material, the effects of duplicity (if any), including the interstellar material around and between the stars, and any and all other effects giving rise to flux originating at or in the vicinity of the star or other object. The star's image is not a point source, and adverse effects can arise from not allowing correctly for the point spread function, including the wings of the image. The image may also be contaminated by adjacent stars: duplicity or in a crowded field.

I_b (λ, t, x, y, ...) is the sky background, for which corrections to our observed Is must be made before we have the "pure" I_s value. There are many ways to do this, for both photomultiplier and for CCD photometry, and many of them will be discussed in detail at this meeting. Suffice to say here that this correction process is one that can lead to lack of precision and accuracy. In addition to the "process" itself, we have the problems of variable background (coming from a variable background in space, due to emission regions, absorption regions, crowding of stars in the field, etc) and from effects arising in the telescope (vignetting, etc), from variable sensitivity over the detector, from variable transmission in the filters, etc. For fainter stars, the sky background can even dominate the signal, and that too leads to a decrease in precision and accuracy.

IM (λ, x, y, ...) is the impact of interstellar absorption (or interstellar reddening, as the absorption is wavelength dependent). Such absorption is likely to be present in most observations and must be taken into account. If the reddening law varies in different regions of the sky, as it undoubtedly does, and if the derived corrections for this effect are not well handled, then errors creep in easily. Systematic effects are quite likely to arise.

A (λ, t, x, y, HA, DEC, ...) is the effect of atmospheric extinction, also wavelength dependent. Clearly, the effect is also time dependent, over many time scales. Its determination is not always straightforward. Stebbins is quoted as saying that to determine extinction well, you must spend all your observing time leaving nothing for your program. Probably true. Besides the effects of air mass differences, there are temperature and humidity impacts, inversion level altitude, east-west and north-south effects (and towards-away from a city or mine or whatever), jet trails, light clouds, heavy clouds, and so on, all of which change on varying time schedules.

T (λ, t, x, y, ...) reflects the various things that the telescope and its optics can do to you, including the time and wavelength dependence of the reflective coatings on the optics. There are other effects as well, such as incorrect baffling, vignetting, ghost images, dust, and so on.

P (λ, t, x, y, ...) are any effects produced by the photometer and its elements, optical or otherwise.

F (λ, t, x, y, ...) is the impact of the filter (or other wavelength discriminator) on the throughput. It is our choice of filters (or spectrometer) that gives us our information resolution. The careful choice of filters is critical to success, for many reasons. I will discuss some of these in a second paper at this conference. However, as in so many other items noted here, filters are wavelength dependent in themselves, and that dependence changes with time, temperature, humidity, dust coverage, and other perverse reasons. One must be aware of all of this, and allow for it if necessary. It is necessary. The detailed transmission curve of the filters must be known, and allowed for. Central wavelength, shape, bandwidth, tails, are all important items.

D (λ, t, x, y, ...) is the impact of the detector, a topic well covered at this present meeting for CCD photometry. Problems also can and do exist with photomultiplier or other types of photometry.

O (λ, t, x, y, ...) is the impact of "all other." There are always some of these. Identifying and allowing for them is the key. Here we might well include any effects produced by differing reduction procedures.

3. Conclusion

With all these potential and real problems, it is a wonder that one can do photometry at all. The pitfalls are so many. The time needed to investigate all of them is necessarily large, and not often available. Factors change with time as well, often on a rather fast time schedule, or discontinuously. Is there no hope? Probably there is. After all, good photometry has been done. But there is no way it can occur without careful attention to all the details, not just to getting the data and using a neat reduction package. There is a lot more to it.

Another Look at Open Cluster Photometry: Precision and Accuracy

David L. Crawford

GNAT, Inc., crawford@gnat.org

Abstract. Great care must be paid in photometric work if one is to obtain precise and accurate data. It is only too easy to make accidental or systematic errors that can lead to serious misinterpretations of the data. Several examples illustrating what can happen are given in this poster paper, taken from published photometry of open clusters and of the interstellar reddening properties. It is easy to see that we are not always getting the accuracy we would expect.

1. Introduction

It is easy to make photometric observations. One has access to a telescope, and a photometer, whether it is a CCD or one with a photomultiplier. Given a clear night, one observes the program stars (in an open cluster or whatever) and a number of standard stars, with the "standard" filters and detectors at the observatory. Later, one does reductions in the "standard" way, and produces a table of results. The scientific analysis follows, then publication. But have we always understood what we have done?

2. A Few Examples From the Literature

Let me give a few quick examples, comparing data from different publications, taken from current research that I have been doing on stars in a few young open clusters. These clusters have very reddened O- and B-type stars. Such cluster data are most useful in studies of color magnitude diagrams, in comparisons to stellar evolution, in studies of interstellar reddening and absorption, and in calibrations of the photometric parameters themselves. Clearly, precise and accurate data are needed. What do we find?

In one such cluster, two relatively bright stars (V=9.44 and 8.95) have six published values, plus my own new data. The V magnitude and the B-V values agree rather well, though the scatter in B-V is higher than would be expected for such bright stars. Probably the precision of some of the data is not too good. Five of the U-B values agree rather well with each other (even rather low scatter), but the other two are off by 0.10 magnitude! This is unacceptable photometric accuracy, and it would seriously compromise any conclusions about the characteristic of the clusters or of the interstellar reddening. Clearly, one of the authors has goofed, and the other has reduced his data to the same (erroneous) (U-B) zero point.

Two fainter stars in the same cluster have data by two of the observers. These stars are fainter, and the data are more likely to be adversely impacted by variable background (and other things). They are also highly reddened. Here are the values for these two stars:

V	B-V	U-B		V	B-V	U-B
13.13	1.23	0.08		13.42	0.93	-0.07
13.02	1.32	-0.08		13.52	0.93	-0.29

Clearly, there are problems. And this is photoelectric photometry, with the published typical value of "observational errors," plus or minus a few hundreds at the most. The real problem comes when one tries to use such data for drawing scientific conclusions. Remember too that the faintest (and reddest, in this case) stars in the data set are often the ones that the critical conclusions about interstellar reddening laws hang on.

In another young cluster, we find the following for six different observers (except for me, not the same ones as in the other cluster), from over ten stars in each comparison (UBV data, and standard error for one star), relative to our data:

V	±	B-V	±	U-B	±
0.005	0.016	0.010	0.008	0.004	0.005
-0.011	0.034	0.004	0.008	-0.011	0.017
-0.014	0.014	-0.010	0.006	0.003	0.008
0.009	0.017	-0.007	0.015	-0.093	0.020
-0.013	0.025	-0.012	0.008	-0.079	0.024
0.006	0.017	-0.026	0.010	-0.080	0.019

Clearly, there are differences in precision and accuracy. I have grouped the data into two sets, to highlight the (U-B) differences. What has happened?

There are four different uvby data sets available too for this same cluster. Here the differences of the other three to ours is (for b-y, m1, and c1):

b-y	±	m_1	±	c_1	±
0.010	0.005	0.001	0.009	-0.005	0.009
-0.043	0.012	0.052	0.011	-0.058	0.024
0.066	0.021	-0.020	0.020	0.051	0.029

Again, clearly there are differences in precision and accuracy. Why the problems? There can be many reasons (note the lists of potential pitfalls given above), most having to do with lack of understanding of the basics of photometry. Let me mention at least two of the key reasons that probably have caused most of the problems here: First, there is variable interstellar reddening, especially in the first cluster, as well as probable variable background (emission nebula and crowded field) problems. Second, the choice of suitable standard stars for transformations was poor (no reddened OB stars!) for a few of the observers. In one case at least, the reduction techniques were quite non-standard and probably led to some of the problems, especially when coupled with the lack of good standard stars. There may well have been other problems as well, of course.

It would be easy to go on with more examples and with more discussion of the whys and wherefores, but I think I have made my point. Remember that these observations were all done by "photometrists," not even spectroscopists or astrophysicists or theoreticians!

3. Conclusion

Be careful! Photometry is full of pitfalls waiting to mess up our work and our conclusions.

Photometric Systems

David L. Crawford

GNAT, Inc., crawford@gnat.org

Abstract. Understanding the basic ideas of photometry and of photometric systems are at the heart of doing good photometry. The best equipment in the world, well tuned up and well understood, is not enough. One must still plan a good observing program, choose a photometric system that matches the science, observe well, and handle the data reduction well. Such obvious things are not always done.

1. Introduction

Photometry is undoubtedly the most fundamental of all astronomical observations. It is hard to imagine any research of any nature that does not use photometric results in some way or another. The measurement of brightness and of broad band spectral features in a star or other object's flux output is at the heart of much of what we all do. We all do photometry or use the results of someone else's observations.

Photometry and spectroscopy are very complementary, as are photometry and theory. In fact, "complementarity" is one of the most operative words relative to approaches and to progress in our understanding the universe about us. However, while basically very simple, photometry is very often much misunderstood. To do good analysis, it goes without question that we need accurate and precise photometry.

What we measure in photometry is really an integral of the effect of many things: the flux output of the star or other object, including any shells or disks about it (and whatever else nature may have put there to confuse us), interstellar matter, the Earth's atmosphere, the telescope, the filters, the detectors, and other things as well. All of these are changing with time. We are trying to estimate something of interest to use, perhaps the stellar apparent brightness or some strong feature in its spectrum. We are trying (we hope) to avoid the adverse impact of other things that may well confuse our understanding. The selection of our observing site, our telescope, our detector, and our filters all can impact the success of our estimate of the parameter of interest. Photometric systems have usually been designed to be efficient estimators of what we are trying to understand and to avoid the adverse impact of those things that are most likely to mess us up. We try to chose the photometric "system" that will do the best job for us. Simple in principle, but often difficult in practice.

Having a good understanding of what is going on is critical. There are adverse effects of rotation, shape of the star, emission, double or multiple stars, line shape, spots, flares, shells, disks, the interstellar matter and its uniformity, variable stellar or emission background, the faintness of our object, atmospheric seeing and variability, clouds, filter leaks and cut-off characteristics, temperature effects and anything involved, detector changes, and so on. With all of this, it seems impossible to think of doing precise, not to speak of accurate, photometry. In addition, there are issues such as separation of the observed parameters, sensitivity to the feature we are trying to measure, the ease of doing it, standard stars and calibrations, reduction methods, information resolution, and so on. How can one manage? But it has been done. But not as often as one would like, or even as one thinks.

2. Real Life in Today's Observing

Let me elaborate a bit more on the real life difficulties of doing photometry that most of us face. Let me call it "Second Class, First Class Photometry," or perhaps "Photometry at a National Observatory." There is a good deal of excellent equipment, including state-of-the-art photometers. However, they are continually being upgraded or improved or worked on, both hardware and software. At each observing run, one is faced with a new setup, hence a new system really. Indeed, each run one may well have a different detector and a different filter set as well. Then there is the problem of instrumentation changes, with all the associated potential for problems (such as wiring mixups). Usually too, there is the fact that one may not get the number of nights needed (if one in fact gets any) and that some of what one gets is cloudy or marginal seeing. Furthermore, there are either no small telescopes anymore or they are disappearing. Hence, adequate time for standards and for understanding systems is impossible. At some of these places, stars and optical astronomy and photometry are certainly not where the action is.

All of this can and does lead to what I have been calling "Second Class, First Class Astronomy." That means that the scientific program is first rate, but the observational restrictions limit what can be done, often severely compromising the effort. I don't think that it is getting any better with the new technology. Things are more complicated, change more often, and many observers are getting further and further removed from the sky, buried in the computer room, and out of touch with both the sky and the instrumentation and the details of the software. Hence, real understanding of what is going on is difficult or impossible. Photometry like that mentioned above may well become more common rather than less common. We must do better rather than worse!

3. What To Do

I won't go into detail here about "What To Do?" Books and papers have discussed this topic at length, and IAU Commission 25 (Photometry) meetings and symposia have dwelt on the issues for decades. I talked at some length about it in a paper at the recent meeting in Cape Town honoring Alan Cousins on his 90th year. It is interesting to note that even today (at least until he passed the age of 90) Cousins is probably doing the most accurate photometry of any of us, and he is doing it from the middle of a large city and with an electron tube amplifier and a Brown strip chart recorder. Sufficient to say here: If you want to get good data, yourself or from the literature, you must understand photometry and photometric systems. They appear like simple issues, but they can be a trap, leading to very interesting but erroneous conclusions about stars, gas, and dust in the galaxy. I give just one example in a poster paper elsewhere in this volume, the data being taken from published papers in the astronomical literature.

We must insist, in astronomy, that a balanced program of support will include support for photometry, for small telescopes, for adequate coordination of their use, and for sufficient time on such telescopes to insure "First Class, First Class Photometry." As I said in South Africa, we could even have a new generation of photometry, with the new panoramic detectors and with an implementation of a global network of small telescopes doing excellent photometry everywhere. If we don't do these things, then I fear that we will face losing touch with observations and the systems. We will see more "second class" observations and science, more systematic errors, more marginal data. Let us hope that we do better in our support of these fundamental things, like standards, photometry, and effective small telescopes.

I think that we can, but we must actually do it. Let me repeat the last paragraph from the paper at the Cousins' meeting, summarizing some of the positive results of such efforts:

"We would then see:

1. New technology, detectors and instrumentation, new telescopes, including new generation small ones, an essential component in a balanced funding approach to astronomy. We need it all, and we will be getting some of it without question.

2. Better understanding of photometry and photometric systems.

3. More and better data. And adequate archiving of the data.

4. More and better standard stars. More tie-ins.

5. A wider range of standard stars, in character and faintness.

6. Better astronomy, better science. Precise and accurate photometry is fundamental to it all".

4. Standard Stars and Standard Systems

(This section is largely extracted from a paper given at a photometric conference in Moletai, Lithuania, in August 1995.)

We all know that photometry is fundamental to progress in almost all astronomical research. It is hard to imagine doing much without a measure of brightness or color or other photometric parameters. To be able to do effective research programs, it is essential to be able to compare the photometric data used with other such data. Hence, we all use standard photometric systems, such as UBV, or uvbyβ, or Geneva, or Vilnius standard systems.

It is not always clear to the user how such standard systems have been defined or are to be matched, or even how to be used. Is it by the careful choice of filters and detector that match the system? Perhaps, but that can be very difficult if not impossible. We don't use the detectors that were used to establish the system originally, for example. Is it by careful choice and observation of a subset of the standard stars that were used to help establish the system? Sometimes, but sometimes not. Sometimes not for a good reason: they are all too bright for our current telescope and detector system, or there are no standards of the stellar type we need. Sometimes for a bad reason: not everyone has the concept of interpolation well in mind, or we don't have the time to do an adequate job.

In some sense, viable photometric systems are rather like the MK spectral classification system: They depend a great deal on the usage of standard stars. What makes a system good is in many ways dependent on the quality and quantity of the standard stars used to help define the system. We can (and should) do the best we can in the choice of spectral wavelengths and bandwidths for the filters and in our choice of detectors and reduction procedures, but the final quality and value of the system will depend a lot on the quality of the standard stars used to help define the system and on the choice of standards used for any observing program with the system. Such standards must span the range of stellar types (or spectral types) that will be observed with the system or on the specific observing program. One must interpolate over all the relevant parameters. Naturally, the standards must be observed and tabulated with high precision and high accuracy. Many examples exist in the literature of problems that arise when these obvious guidelines are not followed. Considerable effort should be made by all photometrists to insure that an adequate set of standard stars exist for the most used standard photometric systems.

Let me state some of the essential precepts as I see them:

1. No useful photometric system can exist without an excellent set of standard stars. While it is important to match the hardware and software criteria as closely as feasible, it is not adequate. The systems are really defined by observations on the set of standard stars, just as the MK spectral classification is so defined. It is the MK Process. See the ASP Conference Series Volume 60 proceedings for excellent discussions of the issues.

2. Such standards must be of such a range in characteristics so as to allow interpolation to always be used. These characteristics include: position on the sky (including northern and southern hemispheres, of course), spectral type and luminosity class, chemical composition (the "third" parameter), magnitudes and color indices, and even rotational velocity, magnetic field, duplicity, and all other characteristics to be expected in "unknown" (= non-standard) stars of the observing program. Naturally, if the objects in question are not stars, then standards of such objects are needed as well.

3. It is essential to understand one's natural system and its relation to the standard system; but it is even more essential to always interpolate. It may not be useful (or as accurate) to use a very wide range of standard stars for a research program on stars with a considerably narrower range in characteristics.

4. Enough standards must be observed to insure that such interpolation is possible and so that enough precision can be obtained. It is important to be able to estimate the observational precision and accuracy, and to obtain also "control" observations as a check on such accuracy estimates.

The above discussion about why one should be using standard stars makes it rather clear what sort of standard stars should be used. It is essential to always interpolate. After all, photometry is really a process of interpolation. All our observing and all our reduction techniques must be designed to facilitate this process. Any trace of extrapolation will usually lead us badly astray.

Hence, one must select an adequate number of standards of the needed characteristics to insure a good job of interpolation in all phases of the observation and reduction process. This may not be easy to accomplish, especially if we are observing either a wide range of "unknowns" or types of objects for which few if any standards exist. In the first case, a rule may be "the more the better." In the second case, one may have to try to develop substandards of the type of characteristics needed.

In the observing itself, one must observe so as to interpolate. This means that one must be observing standards at somewhat higher air masses (and lower) than any of the unknowns. One must start and end the night with standards, and intermingle them throughout the night. Particularly on "all sky" programs, it is helpful if one can randomize the intermixing of standards of different characteristics, so that no pattern is built in that may later cause systematic problems.

Careful and adequate time spent planning the observing program is essential, and will always pay off in increased accuracy. So will adequate time spent observing standard stars, even in the face of time pressures (not enough nights assigned to the program, equipment problems, clouds, and learning curve time). Good equipment (stable and well understood) is essential as well. And a most valuable asset is a quality observing site.

We will then be on the road to achieving precise and accurate data, that is, useful data. We will also have done a great deal in the process of understanding our natural system, or equipment and software, and the transformations to the standard system.

In some sense, it seems impossible to be able to do good photometry. There have been a number of excellent investigations of the inherent problems, many by Andy Young for example. It is easy to find in the literature quite a few examples of less than satisfactory photometry. Fortunately, one can also find a number of examples of very high quality, accurate photometry, on a number of systems. Most of these have followed the precepts noted above very well. It is nearly impossible to do good photometry without doing so.

In the paper given in Mexico City (Crawford, 1994), and in a few earlier papers, I discussed some of the problems facing us in the future:

1. Getting adequate observing time to do our programs, and to observe standards (and other calibrations). The excitement of the field and the increasing number of astronomers (in spite of the funding problems we all face) coupled with the fact that some of the telescopes useful for photometry are actually being closed down means that the pressures for time will only increase.

2. Hence, the time available to investigate standard systems, to continue our efforts to establish more standards (especially ones of additional characteristics, including fainter ones of all types) will be sadly lacking.

3. Equipment is getting more powerful but more complicated. We are getting separated even further from the sky than we have been in the past. It will be harder to understand the content and the performance of our black boxes, both hardware and software ones.

4. The drive to work fainter and to observe strange and wonderful objects means that the extrapolation aspects of photometry may be more in evidence than good astronomy requires. Even photometrists get off the track, imagine what others will be doing.

So what can we do, we who are interested in quality photometry and quality photometric systems and standards? Certainly, we must keep talking up the issues, doing the very best we can in our own programs, insuring that the data being archived is of good quality, and all such necessary and obvious things that we all have been doing in the past. Can we do more?

I think so. One such thing would be to increase (greatly) the number of high quality small telescopes at existing observing sites worldwide, and to use some non-negligible amount of the time on them for standardization efforts. In fact, a number of us have proposed just such an organization: GNAT, a Global Network of Small Astronomical Telescopes. GNAT has been incorporated as a non-profit organization. There are already a number of universities and individuals who are formal members. We expect more in the near future. One or more GNAT telescopes should begin operation this year, and a proposal is being prepared for additional ones at other locations.

GNAT's goals are to be a catalyst for all those interested in small telescopes as viable tools in astronomical research and in science education. A number of papers have discussed the needs for and the potentials for such a global network, including ones at this present meeting.

There is no doubt that GNAT would make a most powerful contribution to the solutions needed relative to standard systems and standard stars. One can conceive even "starting over" in establishing totally new systems designed especially for CCD imaging photometry, but useful for any other photometry. Adequate time would exist to test detectors, filters, and hardware so as to insure that a quality system (close to existing systems, of course) would be established, if a new system was to be contemplated. If not, adequate time would exist to thoroughly test most all our existing systems. One would want to do this in any case, of course. There would also be adequate time to observe a well selected list (a large one) of standard stars of all the needed characteristics. And finally, even adequate time would be available for many existing and future photometrists to work on their research programs.

5. Summary

Photometry has always been of the highest value in astronomical research. It always will be. We must do all that we can to insure that such photometry is of the highest accuracy possible. The existence of well chosen standard systems and standard stars is one of the most powerful tools to insure that reality.

References

Crawford, D. L. 1994, Revista Mexicana del Astronomia y Astrofisica, Vol. 29, 115
Johnson, H. L., & Morgan, W. W. 1953, ApJ, 117, 313
Johnson, H. L., & Harris, D. 1954, ApJ, 120, 196
Morgan, W. W., Harris, D. L., & Johnson, H. L. 1953, ApJ, 118, 92

Atmospheric Constraints on Precision Photometry

Ronald J. Angione

Astronomy Department, San Diego State University and Mount Laguna Observatory, San Diego, CA 92182-1221, angione@mintaka.sdsu.edu

1. Introduction

Variations in atmospheric transmission set important constraints on both the usable wavelength range and observing procedures for precision photometry. The three components of atmospheric transmission are (1) Rayleigh scattering, (2) dust (or Mie) scattering, and (3) molecular absorption. Atmospheric transmission measurements were made at San Diego State's Mount Laguna Observatory (MLO) on 874 days between 1979 and 1993 as part of an NSF funded project on monitoring the solar constant. This 14 year long data set covers both quiescent and perturbed (e.g. volcanic eruptions) atmospheric conditions. The data set is both large enough and sufficiently long to set the natural levels and variations in the components of atmospheric transmission.

2. Observations and Reductions

These data were obtained at MLO using a filter wheel radiometer. The radiometer actively tracked the Sun (one arcminute) from 0800 to 1100 PST making a complete set of measurements every 6 minutes. The optical depth was obtained from the slope of the log (intensity) vs. airmass relation. Precision measurements require a highly stable instrument. The design of the radiometer is similar to a stellar photometer. Sunlight entered through a 5 mm aperture placed behind a quartz window. A quartz Fabry lens imaged the aperture on a EG&G UV 444B silicon photodiode detector. In front of the detector was a 12 position filter wheel.

The radiometer was encased inside an insulated outer box and was actively temperature controlled year round at 38 ± 0.5°C; the detector was controlled at 40 ± 0.1°C. Much of the electronics was also contained in the radiometer and was thus also temperature controlled. The filter wheel contained eleven interference filters (FWHM nominally 75 Angstroms) strategically placed from 3835 to 10097 Angstroms to yield ozone from the Chappuis band, water vapor from the 9350 Angstrom band, and the dust component from the remaining filters. Results show that the extinction can be determined to better than a milli-magnitude/airmass, and that a properly designed and maintained photometer, at least with a silicon detector, can achieve a precision significantly better than a milli-magnitude.

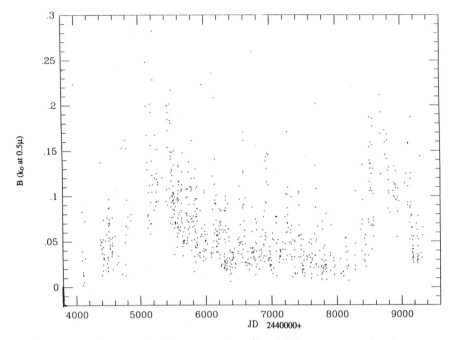

Figure 1. Dust extinction verses time (Rayleigh component has been subtracted). The units of B are the extinction at 1micron.

3. Rayleigh Scattering

Rayleigh scattering is reasonably well understood and well behaved, and is for the case of particles, usually molecules, small compared to the wavelength of light. Let I_o be the incident intensity, τ the optical depth, and X the airmass, then:

$$I = I_o e^{-\tau X} \qquad (1)$$

where

$$\tau = \int_0^X \beta dh \qquad (2)$$

The decadic extinction coefficient $k = 1.086\tau$. For a mixture of gases,

$$\beta = \frac{32\pi^3}{3N^2\lambda^4} \sum_i v_i (n_i - 1)^2 \frac{6 + 3\rho_i}{6 - 7\rho_i} \qquad (3)$$

n is the index of refraction, N is the number of molecules/unit volume, v_i the fractional volume for the i-th gas, ρ_i the depolarization factor, and λ the wavelength. Since the composition, and hence all the variables above, change with altitude, Equation 3 must be integrated through the atmosphere using a program such as LOWTRAN (Selby and MaClatchey, 1972), U.S. Standard Atmosphere

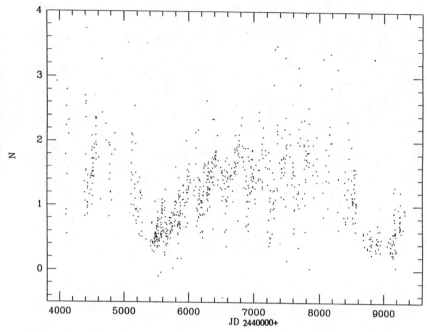

Figure 2. How dust particle size and composition change with time; smaller N means larger particles.

1976 (NOAA-S/T 76-1562), or the European 5S code. Such calculations are tedious, and I have found that the approximation developed by Frohlich and Shaw (1978), with the correction factor of 1.031 (Young, 1980) is sufficient for extinction calculations:

$$\tau_R(\lambda) = (1.031)(\frac{P}{1013.26})0.00838\lambda^{-(3.916+0.074\lambda+\frac{0.05}{\lambda})} \quad (4)$$

Extinction due to Rayleigh scattering scales well with atmospheric pressure, which is easy to monitor at any observatory.

4. Dust Scattering

Gustav Mie in 1908 found a solution to Maxwell's equations for the scattering of light by spherical particles with a diameter comparable to the wavelength of the scattered light. In the real atmosphere Mie, or dust, scattering is too complicated to calculate from basic principles because the particles have (1) a variety of shapes, (2) a variety of composition (e.g. silica, hydrocarbons, volcanic products, etc.), and (3) a distribution of sizes. Empirical studies have found the distribution of sizes is approximately Log- normal.

$$\frac{dn(r)}{d\log r} = Ar^{-m} \quad (5)$$

where r is the radius of the particle. In this case the dust extinction component is:

$$k_D = B\lambda^{-N} \qquad (6)$$

B gives the dust burden, and is numerically the extinction coefficient at 1 micron. N is related to the particle size (m = N+2, in equation 6), and is normally about 1. For particles large compared to the wavelength, N goes to zero, and for small particles N goes to 4 (Rayleigh scattering). For a broad spectral region N must clearly be a function of wavelength; a given particle size will be large for short wavelengths and small for long wavelengths. Following the eruption of El Chichon Equation 6 would frequently no longer fit the data; in a log-log plot of extinction verses wavelength there was pronounced curvature. I then adopted the following empirical formula, which works reasonably well:

$$k_D = B\lambda^{-(N+C\log\lambda)} \qquad (7)$$

Figures 1 and 2 show B (adjusted to the dust component of extinction at $.5\mu$) and N for the period 1979-1993. The eruption of El Chichon (1982) and Pinatubo (1991) are clearly visible as dramatic changes in both extinction, B (higher), and N, smaller (larger particle size). The extinction is also highly variable with a remarkable absence of low values. Note how the particle size appears to grow (N decreasing to zero) for about 9 months after the eruption. This is thought to be volcanic SO_2 slowly forming larger sulfate particles, which gradually reach a size to be optically active and peaking about 9 months after the eruption. From the figures you can see that it takes about 3 years for the particles to settle out of the atmosphere.

Figure 3 shows the extinction during the quiescent period between the two eruptions. The seasonal variation is apparent, as is the fact that extinction can change significantly night to night, and by implication also during the night. Figure 4 shows that the seasonal variation is not well behaved in that the extinction peaks at different times in different years, making the use of mean extinction a risky business.

5. Molecular Absorption

Figure 5 shows the total ozone measured above MLO. It is clearly variable and has a strong seasonal variation. However, the Chappuis ozone band, extending from λ 5000-7000, is a weak, nearly continuous structure absorption feature, and can be treated in the standard extinction reduction method. Its total contribution to the atmospheric extinction is on the order of a few hundredths of a magnitude. This feature "flattens" out the wavelength dependence of extinction across the V-band, normally negating the need for a second-order extinction coefficient.

The water vapor feature shown in Figure 6 was obtained at MLO with a celeostat-fed spectrograph (for a complete description see Angione, 1987). It is a strong, complex feature and can not be readily corrected out of the photometry. Figures 7 and 8 show that it has a well behaved seasonal variation peaking in the late, local summer. Both the monitoring of the water vapor and correcting

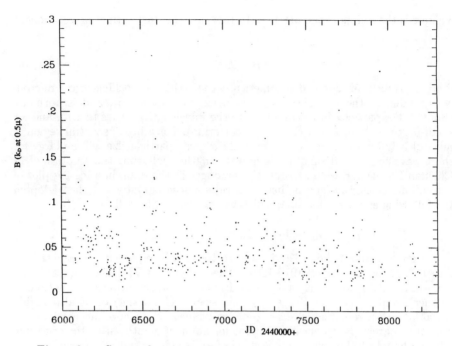

Figure 3. Seasonal variation in dust extinction during an atmospherically quiescent time.

for it are complex. Nor does the transmission in the water vapor band change linearly with airmass.

With a broad band I filter, such as the RG9 used by some observers, the long wavelength cutoff is set by the CCD QE convolved with the water vapor dominated atmospheric transmission. The temperature dependence of QE of the CCD at the red end means that the spectral response of the I filter will also depend on the operating temperature of the CCD. Variations in the water vapor will significantly affect the effective wavelength of the I-band (up to several hundred Angstroms).

6. Conclusions

Precision photometry requires that atmospheric transmission be monitored frequently throughout the night. Special care will be needed in both the measurement of atmospheric extinction and the application of the extinction correction to achieve a precision approaching a milli-magnitude. The precision of the photometry will likely be degraded during atmospherically perturbed times, such as for several years following major volcanic eruptions. The strong influence of the water vapor absorption in the region of the I-band filter may necessitate (1) its exclusion, (2) using an interference filter for the I-band (excluding as much of the water vapor as possible, or (3) developing a special reduction procedure. All these atmospheric problems are especially important to the GNAT

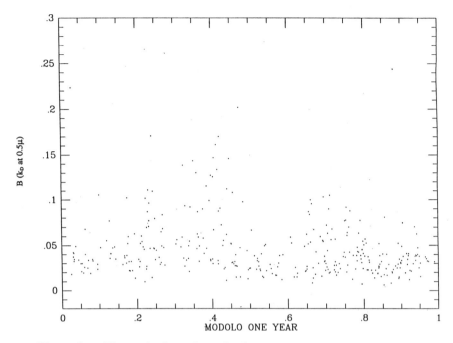

Figure 4. Figure 4. shows how the dust component peaks at different times in different years.

project because of the need to combine photometry from a number of different observatories.

References

Angione, R.J. 1981, in "Variations in the Solar Constant", NASA Con. Proc. 2191, S. Sophia, Ed.

Angione, R.J. 1987, P.A.S.P. 99, 895.

Angione, R.J., and de Vaucouleurs, G. 1986, P.A.S.P. 98, 1201.

Angione, R.J. and Roosen, R.G. 1983, J. Climate and Appl. Met. 22, 1377.

Frohlich, C. and Shaw, G. 1978, Phys. Meteorol. Obs. World Rad. Center Pub. No. 557.

Selby, J.E.A. and MaClatchey, R.M. 1972, AFCRL-72, Environ. RES Papers No. 427.

Young, A.T. 1980, Appl. Optics 19, 3427.

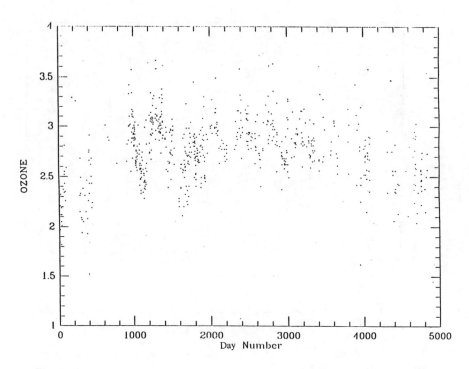

Figure 5. Seasonal and short-term variations in the total ozone level.

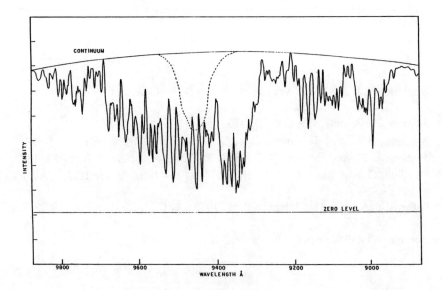

Figure 6. The 0.935 micron water vapor band that lies in the I-filter.

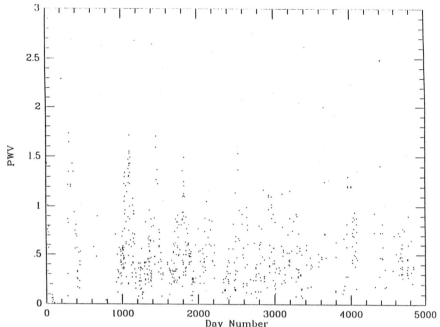

Figure 7. The variation in precipitable water vapor (PWV) showing the pronounced seasonal variation.

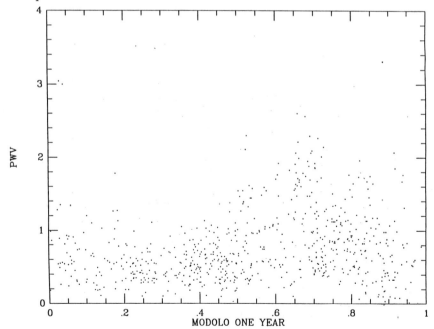

Figure 8. The well-behaved season variation in precipitable water vapor peaking in late summer.

Some Practical Aspects of CCD Camera Construction

Roy A. Tucker

National Solar Observatory, Tucson, AZ 85719 USA

1. Introduction

Precision CCD photometry requires a camera system with known and stable characteristics. In recent years, the CCD imaging devices and the electronics that support them have become very good, asymptotically approaching perfection. Careful consideration must be given to other aspects of the camera design such as device refrigeration and mechanics to avoid compromising this high level of performance. For the benefit of those who have little previous experience in camera construction, I would like to introduce and briefly discuss some of these issues.

2. Device Refrigeration

As seen in figure 1, silicon CCD imaging devices experience a dark current that is a function of temperature. This graph shows characteristic curves for six different levels of dark current generation ranging from 10 to 5000 picoamps per square centimeter at 293 Kelvins. These curves indicate the dark current in electrons per pixel per second (assuming 24 micron pixels) versus temperature in degrees Kelvin.

This thermal dark current has a shot noise component that is equal to the square root of the total accumulated thermal charge and will quickly exceed the read noise of the detector if it is not controlled. Although some techniques, such as Multi-Phase Pinned (MPP) architecture, have been developed to reduce the thermal dark current, silicon CCD imagers in most astronomical applications are refrigerated to reduce this source of noise.

How much the CCD should be cooled depends upon the intended application. In asteroid or supernova search programs where the imager will be coupled with a relatively fast optical system and operated without color filtering, the background skyglow will be the primary limiting factor. Only a moderate degree of cooling will be required to lower the dark current to a level much below the skyglow. On the other hand, observations of faint objects with a high-dispersion spectrograph will result in very few photons per pixel and the detector should be quite cold.

Figures 2, 3, and 4 show that the quantum efficiency of a CCD may be reduced when operated very cold. CCD imagers should be carefully characterized to determine the degree to which they are affected by cooling. Situations requiring the greatest possible sensitivity, especially near the red end of the CCD response, will require a careful choice of detector temperature to balance quantum efficiency against dark current when using a device that is subject to this effect.

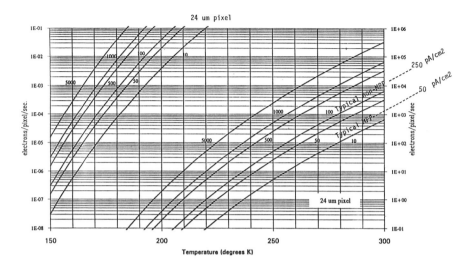

Figure 1. Effect of temperature on dark current. Parameter is pA/cmsq at 293 K. (Courtesy of Scientific Imaging Technologies)

3. Methods of Refrigeration

There are many methods available for providing the cooling required for a particular application. Commonly available cryogens are liquid nitrogen (77 degrees Kelvin) and dry ice (195 degrees Kelvin). The Stirling cycle cooler is a popular mechanical refrigerator that provides temperatures in the range of 60 to 150 degrees Kelvin. The Joule-Thomson cooler uses the expansion of high pressure gas to produce temperatures at the boiling point of the gas, 77 degrees Kelvin for nitrogen. Thermoelectric coolers, sometimes known as Peltier devices, are electrically-powered, solid-state refrigerators that can provide modest cooling with great ease and low cost.

4. Cryogenic Cooling

The temperature of liquid nitrogen is too low for proper CCD operation. In electronics, a voltage divider network is used to produce a voltage intermediate between two other voltages. To produce the desired CCD operating temperature, the thermal equivalent of a voltage divider may be arranged between the liquid nitrogen reservoir and the walls of the cryostat. The thermal resistance of the heat flow pathway from the surrounding cryostat walls to the CCD support structure is determined and the thermal resistance of the pathway from the support structure to the liquid nitrogen reservoir is adjusted accordingly.

The simple temperature divider construction described above cannot precisely regulate the final CCD operating temperature. The heat flow pathway to the LN2 reservoir may be some fixed and easily adjustable part such as a piece of copper wire. The thermal resistance of the pathway from the CCD support

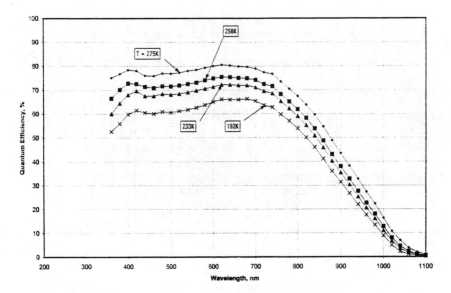

Figure 2. Quantum efficiency vs. wavelength for four different temperatures. (Courtesy of Scientific Imaging Technologies)

to the cryostat walls will be some complicated composite of fiberglass spacers, screws, residual air, infrared radiation, etc. The variation in the temperature of the cryostat walls and the residual atmosphere will lead to variations in CCD temperature. One could provide better thermal control by adding a wire or other easily adjustable thermal link between the CCD support and cryostat walls that would dominate the characteristics of the heat flow between those structures. However, it would also be necessary to decrease the thermal resistance of the CCD to LN2 reservoir link to maintain the desired temperature, resulting in an increased cryogen consumption rate and a reduced hold time. This method is satisfactory as long as the reservoir capacity is large enough to provide a suitably long period between refills of cryogen.

The best method of temperature regulation involves a closed control loop. The thermal resistances are adjusted so that the CCD support structure will equilibrate to a temperature that is too cold. An electrical resistor is mounted on the support structure so that it can act like a heater. A temperature sensor is monitored and electrical power is provided to the resistor to maintain the sensor at a precise temperature.

5. Mechanical Refrigeration

Among the mechanical refrigeration devices currently available, the Stirling cycle cooler has become quite popular in applications such as portable, handheld infrared cameras. A device having dimensions of a few centimeters and weighing a few ounces can convert a few watts of electrical power into cryogenic cold. The technology is very mature and a Mean-Time-Before-Failure of several years is

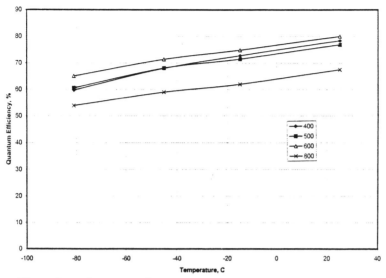

Figure 3. Quantum efficiency vs. temperature for four different wavelengths. (Courtesy of Scientific Imaging Technologies)

often quoted by manufacturers. Disadvantages are mechanical vibration and a high cost of several thousand dollars. However, balanced against the cost of purchasing and storing cryogenic fluids such as liquid nitrogen, this expense might be quite acceptable, especially in applications involving remote, unattended operation. Temperature regulation may be achieved by controlling the speed of the refrigerator.

6. Joule-Thomson Refrigerator

The Joule Thomson refrigerator uses high-pressure gas to create low temperatures. Temperatures equal to the boiling point of the gas may be achieved. A typical device may use 6000 psi nitrogen gas to achieve temperatures of 77 degrees Kelvin. As you might expect, temperature control may be effected by adjusting the pressure or flow rate of the refrigerant gas.

7. Thermoelectric (Peltier Device) Coolers

Currently available thermoelectric coolers use a semiconductor material such as bismuth antimonide in an array of N and P type material. The elements of the array are arranged so that when current is passed through them, heat is pumped from one side of the device to the other. Under ideal circumstances at room temperature, a single such device may develop a difference in temperature of approximately 60 degrees Centigrade. These devices become less efficient at lower temperatures so that at temperatures of minus 60 degrees Centigrade, the same device may produce a difference in temperature of only 20 or 30 degrees

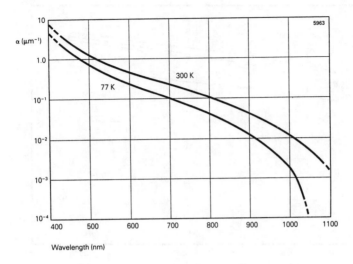

Figure 4.   Silicon optical absorption coefficients. (Courtesy of EEV)

Centigrade. It is possible to construct a multi-stage cooler by stacking a number of devices on top of each other. The heat arriving at the hot side of the device is the sum of the heat pumped from the cold side plus the ohmic heat developed by the passage of electrical current through the device. For this reason, a multi-stage cooler must be constructed like a pyramid, with the smallest stage at the top and the largest at the bottom. Properly designed, constructed, and operated, a multi-stage cooler can produce lower temperatures than dry ice, minus 78.5 degrees Centigrade.

8. Vacuum

A vacuum reduces the flow of heat from the walls of the cryostat to the detector and prevents the condensation of atmospheric water vapor on the detector.

The construction of a cryostat invariably involves a number of seals. The elastomeric "O"-ring seal is popular for seals that will be opened and closed many times. Many different compounds are used in and making "O"-rings. The most suitable for vacuum applications are Butyl rubber and Viton fluorocarbon elastomer. These two substances have the lowest permeability for atmospheric components and have very little in the way way of volatile components that will outgas and contaminate the vacuum system.

The normal construction of an "O"-ring seal involves the placement of an "O"-ring in a recess called the gland or groove and facing a flat surface. The permeability of an "O"-ring seal is affected by the degree of compression and the presence or absence of a vacuum grease. A seal without grease will have the same low permeability with a compression of 50 percent as a seal with grease will have with a compression of 20 percent.

Even the best "O"-ring material is subject to the slow diffusion of air and water vapor resulting in a gradual increase in internal pressure. Only hermetic and hard metal seals are not subject to diffusion. A well constructed Viton "O"-ring seal will typically display a leakage rate of about 2.5 x 10 -8 Torr liters per second per linear inch of seal.

Vacuum Leak Rate Calculations

$$L = .7FDPQ(1-S)^2$$

where:

L = Approximate leak rate of seal in std. cc/sec.

F = Permeability rate of the gas through the elastomer in standard cc/cm^2 sec bar.

D = Inside diameter of the O-ring in inches.

P = Pressure differential across the seal, lb/in^2.

Q = Factor depending on the percent squeeze and whether the O-ring is lubricated or dry.

S = Percent squeeze on the O-ring cross-section expressed as a decimal. (20% squeeze, S=.20)

Seals that are very rarely opened may use some type of hard metal-to-metal seal. Normally copper seals are sandwiched between stainless-steel hardware under considerable mechanical pressure. Once such a seal is broken the copper gasket must be replaced. However, such seals are essentially impermeable and are suitable for high vacuum applications.

Indium metal has a very low vapor pressure and may be used in a variety of vacuum sealing situations. Indium is a remarkably soft metal. Although it melts at approximately 300 degrees Centigrade, it has one of the greatest spans between melting point and boiling point among the pure elements. It also has the remarkable property that it can be used to solder glass. As such it may be used to produce totally impermeable seals between glass and metal. Indium may also be used in compression-type vacuum seals with the added benefit that the indium metal may be reused after the seal is broken.

The vacuum in a closed in cryostat is subject to deterioration due to diffusion of outside air through seals, outgassing, and virtual leaks. A virtual leak is produced when air is trapped in a small internal cavity and slowly seeps out. An example of a virtual leak is the bottom of a screw hole. Perhaps a screw is used to secure a circuit board in place. The compressed circuit board will make a fairly good vacuum seal, trapping air inside the screw hole. After the cryostat is closed up and evacuated, this trapped air will slowly seep out raising the internal pressure. For this reason, vented screws with holes drilled through their length are often used in the construction of a cryostat. Consider also the air trapped in the bottom of an "O"-ring groove under a tight fitting "O"-ring. For this reason "O"-ring grooves should be generous in width to avoid trapping air on the vacuum side of the bottom of the groove even after compression of the seal. Virtual leaks are insidious and will cause great frustration in efforts to achieve good cryostat performance. Give careful thought to your cryostat designs to avoid the possibility of trapping air.

Much outgassing can be traced to two sources: water vapor and more complicated molecules, typically from plastics. For example, polyvinyl chloride may slowly outgas vinyl vapor and hydrogen chloride. For this reason, use of plastics must be minimized in a cryostat. During evacuation of the cryostat, the most common atmospheric components, oxygen, nitrogen etc., are quickly removed. Water vapor adsorbs tightly to any interior surface and will require considerable pumping to remove. Since the amount of adsorbed water vapor is a function of surface area, it is best that the internal surfaces be polished. The removal of water vapor may be facilitated by heating, vibration, ultraviolet radiation, or corona discharge. Table 1 information below is courtesy of Parker Seals.

TABLE I — WEIGHT LOSS OF COMPOUNDS IN VACUUM

Test Samples: Approximately .075 thick
Vacuum Level: Approximately 1×10^{-6} torr
Time: 336 hours (two weeks)
Room Temperature

POLYMER	PERCENT WEIGHT LOSS	POLYMER	PERCENT WEIGHT LOSS
Butyl	.18	Nitrile	1.06
Neoprene	.13	Polyurethane	1.29
Ethylene Propylene	.39	Silicone	.03
Ethylene Propylene	.92	Silicone	.31
Ethylene Propylene	.76	Fluorocarbon	.09
Fluorosilicone	.28	Fluorocarbon	.07
Fluorosilicone	.25		
Nitrile	3.45		

NOTE: Varying weight loss figures within the same polymer family represents different formulations, or compounds, made from that polymer.

While some camera cryostats are continuously pumped, for example, with an ion pump, most are simply sealed off after being evacuated. The increase in the pressure due to diffusion of air through the seals or outgassing of volatile and water vapor can be controlled to some extent by the phenomenon of cryopumping and the introduction of molecular sieve material.

Cryopumping occurs when the cold temperatures produced by the refrigeration cause volatile substances in the cryostat to condense, thereby reducing the interior pressure of the cryostat. Even the modest degree of refrigeration produced by Peltier coolers can condense water vapor and reduce the partial pressure to 10 microns at -58 degrees Centigrade or 1 micron at -74 degrees.

Molecular sieve materials are characterized by having a very large percentage of their volume occupied by voids with dimensions measured in Angstroms. This produces enormous surface areas upon which volatile substances may be adsorbed. Frequently used materials include coconut shell charcoal, zeolites (clay-like substances), and alumina. Alumina has an adsorption surface area of 210 square meters per gram. Coconut shell charcoal has an adsorption surface area of 2500 square meters per gram. There are several types of zeolite molecular sieve materials. Types 4A, 5A, and 13X are commonly used in vacuum applications. Type 13X has an adsorption surface area of 1030 square meters per gram. All of these substances are excellent for the adsorption of water and oil vapor within the cryostat. The adsorptive affinity of these substances is increased with a reduction in temperature. Usually a small container of one of these substances will be attached to the coldest structure within the cryostat. Zeolite molecular sieve materials, when cooled with liquid nitrogen, may be used as a non-mechanical means of generating a rough vacuum in a vacuum system.

A material often referred to as "super insulation" may be used as a multi-layer thermal blanket to further reduce the transfer of heat by radiation and residual air. Super insulation is a thin Mylar film with a metallized coating and a crinkled texture. Although it assists in reducing heat flow within the cryostat, it also provides a large surface area for the adsorption of water vapor, increasing pump down time.

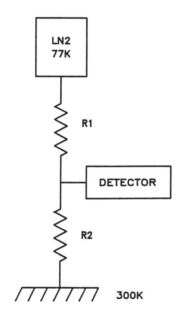

Figure 5. Simple heat flow temperature divider model.

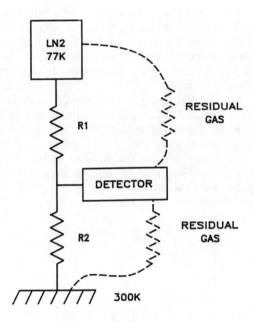

Figure 6.   More realistic heat flow temperature divider model.

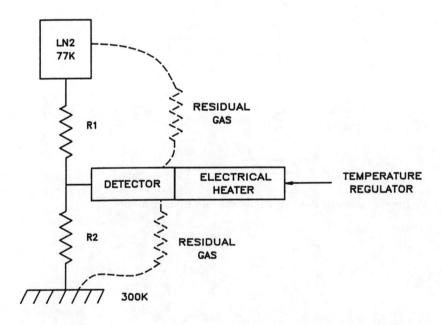

Figure 7.   Electrical temperature regulation of a heat flow temperature divider.

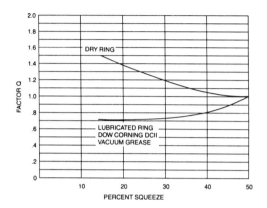

Figure 8. "O"-ring leak rate, dry vs. lubricated. (Courtesy of Parker Seals)

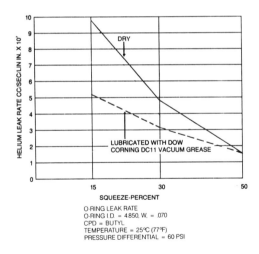

Figure 9. Effect of squeeze and lubricant on "O"-ring leak rate. (Courtesy of Parker Seals)

ELIMINATION OF VIRTUAL LEAKS WITH VENTED SCREWS

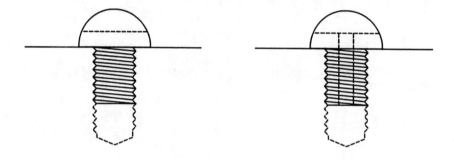

Figure 10. Vented screws may be used to prevent a virtual leak.

Stellar Photometry Tools in IRAF

L. E. Davis

National Optical Astronomy Observatories, Tucson, AZ 85719

Abstract. The current IRAF software tools for planning, acquiring, reducing, and analyzing stellar CCD photometry data are described.

1. Introduction

The current IRAF tools and IRAF documentation used for planning, acquiring, reducing, and analyzing stellar CCD photometry data are described, and related references in the literature are sited.

2. The IRAF System

IRAF stands for Image Reduction and Analysis Facility. IRAF is a software system designed for the reduction and analysis of astronomical data and is written, supported, and distributed by NOAO. The distributed IRAF system includes a core system of general image processing tasks, the NOAO package for the reduction of ground-based optical and infrared data, a scripting facility, and a programming environment. IRAF addons include several NOAO packages, several large addon packages distributed by other observatories including STSDAS, TABLES, XRAY, and EUVE, and a variety of smaller contributed packages. More information about the IRAF system and its support services can be found at the IRAF website iraf.noao.edu/.

3. CCD Observation Planning Tools

Planning a program of stellar CCD photometry observations requires: choosing the program and calibration stars, estimating their magnitudes, selecting the correct combination of telescope, detector, shutter, filter, site, and sky conditions required to achieve the program's scientific goals, and estimating the exposure time required to achieve the desired signal-to-noise.

At the time of writing no IRAF tools for choosing good calibration stars exist. However several optical and infrared standard star photometry catalogs are available on disk (see section 8.4), from which good calibrations stars can be chosen. Calibration stars should have accurate photometric indices in a filter system that is a good match to the observatory system, be bright enough to be accurately observed with short exposures, faint enough not to saturate in short exposures, and cover the expected airmass and color range of the program stars.

The IRAF CCDTIME task can be used to estimate exposure times for stars of a given apparent magnitude. CCDTIME uses empirical data about the telescope, detector, filter, and site conditions, plus user supplied information on the expected seeing, phase of the moon, airmass, and desired signal-to-noise ratio, to estimate the required exposure time. CCDTIME comes configured for the KPNO instruments but users can create database entries for other sites and instruments. A description of CCDTIME can be found in the IRAF on-line help pages and can also be accessed from the web at iraf.noao.edu scripts/irafhelp?ccdtime.

The addon STSDAS SYNPHOT package can be used to predict count rates for program objects given, component throughput functions for all the optical elements, a component configuration table describing how the components are ordered, and approximate magnitude and spectral data for the star of interest. Although SYNPHOT comes configured for the ST instruments, it can be used to predict count rates for any system. Interested readers should consult Horne (1988) and Koorneef et. al (1985) and references therein for background information on synthetic aperture photometry techniques, and Bushouse (1995) for a user's guide to SYNPHOT.

4. CCD Data Acquisition Tools

Acquiring CCD data normally taking exposures, evaluating the incoming data, and writing the data to an archiving medium. Aside from the scientific quality of the data and the efficiency of the data taking process, the data acquisition issues which usually concern users are: ease of use, the availability of quick-look tools, and the readability, completeness, and portability of the archive format. Data formats and quick-look tools are discussed in the following sections. The remaining issues are discussed below in the context of 3 currently active IRAF observing environments: ICE (the IRAF Control Environment), CCDPhot (the CCD Photometer), and MDHS (the Mosaic Data Handling System).

ICE is the primary optical CCD data acquisition system at KPNO. Most ICE commands are a variation of a single OBSERVE task which prompts the user for exposure time and filter, takes the exposure, and writes the results to an IRAF image on disk and a FITS format file on tape. Once on disk various IRAF tools can be used to display and examine the image and perform basic CCD reductions. The latest versions of ICE and IRAF support reading and writing FITS images directly on disk. ICE was originally developed by Skip Scaller of Steward Observatory (Scaller 1992) and modified for use at KPNO by local staff members (Massey et. al 1996).

CCDPhot is an IRAF task which takes CCD images in several filters sequentially and reduces them to photometry records containing instrumental magnitudes in real-time. CCDPhot currently runs at the KPNO 0.9m telescope and is not accepting new proposals, but does provide an interesting example of the kind of real-time application IRAF is capable of supporting. The CCDPhot design and usage are described in Tody and Davis (1992) and Davis et. al (1996).

The MDHS is the newest IRAF data acquisition environment. The current working system features support for real-time image display, on the fly calibrations for better quick-look analysis, a GUI control interface, and writing fully documented FITS files directly to disk. The system architecture is based on general message bus and distributed objects technology. Interested readers should consult the design paper Tody and Valdes (1998), various other MDHS related papers collected at iraf.noao.edu iraf/web/projects/ccdmosaic/, and the latest version of the user's manual (Armandroff et. al 1998).

The IRAF data acquisition packages are not distributed with basic IRAF but are available as addon packages. They can be built and run on any platform which supports IRAF, but the user must provide a small number of host interface routines for communicating with their local hardware. ICE currently runs successfully at several observatories besides KPNO and Steward, and ports of all or part of the MDHS are underway at the CFHT in Hawaii and other sites.

5. Data Format Tools

The fundamental image data format in astronomy is FITS (for more information see ftp:// fits.cv.nrao.edu/fits/ FITS.html). Almost all data acquisition systems must either transform their native format to FITS image format at some point or write FITS files directly. Much of the time this process is transparent to the user but there are a few issues that users should be aware of. Most raw CCD data is in 16 bit signed or unsigned integer format. FITS supports 16 bit and 32 bit signed integers. 16 bit unsigned data must either be written in FITS 32 bit signed format or scaled to the FITS 16 bit signed format. The most likely cause of IRAF or other software failing to read FITS 16 bit data correctly is that it was originally 16 bit unsigned data that was incorrectly converted to FITS. Fortunately it is easy to reformat this data without loss of precision. Reduced CCD data is usually written in 32 bit floating point format, which should be archived to the FITS 32 bit IEEE format. Trying to save space by compressing 32 bit real numbers into 16 bit integers can result in loss of precision and increased image noise.

FITS supports a simple keyword = value syntax for storing information about the telescope, detector, and observing conditions in the image header. Unfortunately there is little agreement on the definition of standard keywords for describing CCD observations and reductions, so many observatories including NOAO have adopted a set of their own. Some data acquisition systems write little or no information to the image header. Both missing and non-standard information can complicate the later reduction and analysis steps.

IRAF has tried to address some of these data format issues by supporting the FITS image kernel. The FITS image kernel permits IRAF tasks to read and write FITS format images directly, and is now part of the IRAF standard system. Interested readers can find out more about this facility in Zarate (1998). NOAO and IRAF staff have also invested considerable effort over the years towards the goal of defining a standard set of CCD instrument keywords, ensuring that the NOAO data acquisition systems write complete and accurate information in the image headers, and encouraging other observatories and instrument developers to adopt a standard keyword set. The latest efforts in this area are documented in Valdes (1996).

6. CCD Observing and Quick-Look Tools

Most observing program require some means of examining and determining the quality of their data as it is being taken. Typical concerns at this point in an observing program are the following. Does the data look ok? Is the observed readout noise consistent with the theoretical readout noise and gain setting? Is the DC offset on the detector reasonable and stable? Is there structure in the bias frames and if so is it stable? Is there a measurable dark current, does it have any structure, and if so is the structure stable? Do the flats contain enough but not too many counts? Is there evidence for transient structures, e.g. doughnuts in the flats? Are there bad regions on the chip that must be avoided? Are dome flats good enough or are twilight and / or sky flats required? Is the information in the image header complete and correct? Is the focus good? Is the seeing good and stable? Is the chip linear over the expected count range? Does the photometric zero point vary significantly from one part of the chip to another? Problems in any of these areas can severely impact the quality of the photometry derived from the data. Excellent discussions of the kinds of calibrations that should be taken and the kinds of checks that should be made at the telescope, can be found in Massey et. al (1997) and Massey and Jacoby (1992) respectively.

In IRAF most basic quick-look functions are provided by the XIMTOOL image display server (the alternate image display servers SAOIMAGE and SAOTng are also available) and the DISPLAY and IMEXAMINE tasks. Image headers can be examined with the IMHEADER and HSELECT tasks. The tasks STARFOCUS and KPNOFOCUS in the NOAO addon NMISC package can be used to estimate the best focus from a single focus frame or a series of frames. The gain and readout noise can be estimated from a set of bias and flat field frames using the FINDGAIN and FINDSHUTCOR tasks. A guide to using the IRAF quick-look facilities in the ICE data acquisition environment can be found in Massey et. al (1996). Parallel facilities for MDHS environment are described in Armandroff et. al (1998).

7. CCD Reduction Tools

Before any photometric measurements can be made the detector signature must be removed from the data. This process may involve one or more of the following steps: DC offset subtraction, DC pattern subtraction, dark-current subtraction, flat-fielding to remove pixel to pixel gain variations, flat-fielding to remove large-scale gradients, cosmic ray removal, and combining images to improve signal-to-noise, remove gaps in the observations, or increase the field of view. The goal of the reductions is to remove the detector signature while maximizing the signal-to-noise and precision of the data.

The CCD reduction procedures should introduce as little additional noise as possible, preserve the quality of the photometry, and preserve the image statistics. For this reason users should normally take several bias and flat-field images and carefully combine them so as to beat down the noise in the calibrations. Although raw CCD data is usually in integer format the reductions should be done and written out in floating point to avoid truncation errors. Detailed discussions on modeling the noise in CCD data can be found in Newberry (1991), Howell (1992), and Shaw and Horne (1992).

For precision photometry work flat-fielding is usually the most critical step. In order to achieve 1% all sky photometry the images must be flattened to 1% or better. A good discussion of flat fielding issues can be found in Massey et. al (1997) and Massey (1997). Users taking data with wide field instruments such as the NOAO mosaic may also need to consider the affects of a variable pixel scale on their reduction and analysis procedures. For images with variable pixel scales conventional flat-fielding techniques will result in a variable photometric zero point which must be dealt with in the subsequent analysis procedures. For a good discussion of these issues see Armandroff et. al (1998).

Cosmic ray removal is usually done by splitting exposures into several parts, and then recombining the images using a rejection algorithm to remove the cosmic rays. This technique works well for long exposures with low readout noise chips. Removing cosmic rays from single images is more difficult and requires training the software to distinguish cosmic rays from stellar objects or removing them with some kind of filtering process. If users are going to do PSF fitting photometry of stellar objects, it may not be necessary to remove cosmic rays, as bad pixels can be automatically detected and removed from the fit.

In IRAF basic CCD reductions are done with the CCDRED package. CCDRED is a non-interactive header driven package which can successfully reduce date taken with many different instrument, detector, and observatory combinations. Users taking data with non-NOAO instruments must supply the package with a translation table which tells CCDRED how to transform "foreign" keywords to the expected NOAO ones. CCDRED is described in Massey (1997). MSCRED is the CCD mosaic reduction package. MSCRED is very similar to CCDRED but includes additional routines for combining the mosaic pieces and dithered mosaic observations (Armandroff et. al 1998). Like CCDRED, MSCRED can be used for non NOAO data given suitable keyword translations. Other CCD instrument reduction packages can be found in the external addon packages, e.g. the WFPC package inside STSDAS. A new NOAO addon package for removing cosmic rays CRUTIL is now available.

8. Stellar CCD Photometry Tools

There are many approaches to analyzing stellar photometry data. This section outlines a set of "standard" procedures for analyzing multi-filter, ground-based, all-sky stellar CCD observations, where the observations consist of a large number of repeated short exposures of standard stars in uncrowded fields, and several long exposures of crowded program fields. Detailed descriptions of the IRAF implementation of these procedures with examples can be found in Massey and Davis (1992) and Davis (1994). For a recent example of published science produced using these procedures within IRAF see Massey (1998).

8.1. Preparing the Data for Analysis

Before computing a single magnitude users should review the observing logs and reduction logs if any, and check for records concerning photometric conditions, seeing changes, instrument problems, or reduction problems that might impact their analysis procedures.

Users should check that the following items are present and correct in the image headers and add them in if they are not: the gain and readout noise of the detector, the exposure time in a consistent set of units, the filter id, the airmass, and the time of the observation in a consistent set of units. A consistent set of exposure times is required to normalize the computed magnitudes. The filter, airmass, and time of observation are not required for making photometric measurements, but their presence will greatly simplify the later calibration steps. The gain and readout noise are required for computing accurate aperture photometry magnitude errors, and in the case of the PSF fitting code, for computing optimal weights, identifying bad points, computing correct magnitude errors, and computing meaningful chi values. Users should also determine the high and low linearity limits for their detector. These good data limits are used to flag bad aperture photometry measurements, and to reject bad data from the PSF fitting code.

In order to achieve the highest precision photometry possible it may be necessary to correct the image header exposure time for shutter error, the airmass to mid-exposure, and the time of observation to mid-exposure. Shutter corrections are normally only important for short exposures, and airmass and time of observation corrections for long exposures. If the image to be reduced is the sum, average, or median of several images the effective gain and readout noise must be computed from the individual image gain and readout noise values and the number of images. As a check of the image statistics users should compare the measured standard deviation of the pixels in a blank sky region, with the value predicted from the known effective gain and readout noise values and the average sky value. If the predicted and observed sky noise do not agree, as may be the case for example if a mean background level has been subtracted from the image, the source of the discrepancy should be tracked down and its possible impact on the analysis procedures understood before proceeding. Details of how to compute the effective gain and readout noise and perform the sky noise test can be found in Davis (1994).

The IRAF DISPLAY and IMEXAMINE tasks and the XIMTOOL image display server can be used to check the accuracy of the flat-fielding and other calibrations and to assess the seeing throughout the night in question. Basic image header examining and editing functions can be performed with the IMHEADER, HSELECT, HEDIT, and ASTEDIT tasks. The SETAIRMASS task can be used to compute the effective airmass and exposure time.

8.2. Finding the Standard Stars

In most cases the user must supply pixel coordinates for the standard stars to be measured, either interactively via the image display, or via a pixel coordinate list. If the celestial coordinates of the standard stars are known sufficiently accurately, and the image has an accurate world coordinate system in its header, pixel coordinate lists can be created automatically from the celestial coordinate lists using the header information. If the standard stars are on average brighter than neighboring stars in the field, then an automatic star finding procedure can be used to generate a pixel coordinate list.

Any program which writes a text file with x and y pixel coordinates in columns 1 and 2 can create a pixel coordinate list suitable for input to the IRAF photometry software. Image cursor pixel coordinate lists can be generated with the RIMCURSOR task. Celestial coordinate lists can be transformed to pixel coordinates lists using the IMMATCH package tasks WCSCTRAN and SKYCTRAN. In most cases pixel coordinates generated in either way must be recentered by the photometry code before accurate photometry measurements can be made. Pixel coordinate lists can also be created by the automatic star finding task DAOFIND in the NOAO packages APPHOT and DAOPHOT. The DAOFIND coordinates are computed using an accurate 1D Gaussian centering routine and do not need to be recentered. Details about the star finding algorithm used in the APPHOT and DAOPHOT packages can be found in Stetson (1987), and a discussion of its IRAF implementation in Davis (1994).

8.3. Measuring the Standard Stars

Standard stars are usually bright objects in uncrowded regions and thus can be measured with standard aperture photometry techniques. Good discussions of the theory and practice of aperture photometry can be found in Stetson (1987), Massey et. al (1989), and Howell (1989). The basic measurement technique involves: reading an initial x and y coordinate, choosing a centering box and computing new x an y coordinates, choosing a sky annulus and measuring the sky background value using pixels inside the annulus, choosing an aperture radius and measuring the number of pixels and counts inside that radius, subtracting the sky value from all the pixels in the photometry aperture, and computing the magnitude. Users interested in alternative aperture photometry techniques for uncrowded regions should check out the weighted optimal extraction algorithm described by Stover and Allen (1987), although it appears to require well-sampled data to be effective (Abbott et. al 1992).

If recentering is required, the centering box should be set to 2.0 - 3.0 FWHM to get achieve good results. For bright objects measured through large apertures, the choice of centering algorithm is not critical, and simple 1D centroiding will normally yield 0.02 - 0.10 pixel accuracy. For the fainter stars 1D Gaussian centering may yield more accurate results than simple centroiding, and for the faintest objects 2D algorithms will do better still. Interested users can find more information on digital centering algorithms in Chiu (1977), Goad (1986), Stone (1989), and Lasker et. al (1990) and references therein.

The sky annulus should be placed far enough from the star to avoid contamination from the star itself, but close enough to accurately represent the light under the star, and be large enough to contain enough pixels to yield good measurement statistics. For typical ground-based observations where FWHM ~ 3 pixels, the sky annulus should start several pixels from the edge of the photometry aperture, be no closer in than ~ 6 FWHM from the center of the star, and be wide to contain several hundred sky pixels, ~ 5 pixels wide is usually plenty. There are many ways to estimate the sky value but most algorithms compute some kind of median or mode statistic after rejecting deviant pixels. In relatively uncrowded regions with well-behaved statistics, the choice of algorithm is usually not critical. Discussions of sky fitting algorithms can be found in Adams et. al. (1980), Stetson (1987) and Lasker et. al (1990).

In most cases the aperture radius chosen to measure the standard stars should be suitable for measuring the entire night's data in all filters. For high signal-to-noise measurements the aperture size is not critical; it should be large enough to avoid PSF, seeing, and guiding errors, but not so large that the odds of including bad pixels in the measurement aperture increase dramatically. For interesting discussions of how big a star really is and why aperture photometry works at all see King (1971), Massey et. al (1989), and Massey and Davis (1992). Experience at KPNO suggests that for typical ground-based data with FWHM ~ 3 pixels, the aperture radius should be around 15 - 20 pixels. A more rigorous method of determining the optimal aperture radius involves measuring the standard stars through several apertures, e.g at 5, 10, 15, 20, 25 pixels, up to some large aperture, e.g. 25 - 30 pixels, and computing the magnitude differences between the smaller apertures and the larger one. The smallest aperture for which these differences become constant to within the expected noise is the optimal measuring radius.

The IRAF multi-aperture photometry tasks are QPHOT and PHOT in the APPHOT package and PHOT in the DAOPHOT package. QPHOT is a simpler version of the more flexible but complex PHOT task. Users who plan to do PSF fitting photometry of their program fields should use the PHOT task in the DAOPHOT package; users planning to do aperture photometry measurements only should use the version of PHOT in the APPHOT package or the QPHOT task. Various tasks in the LISTS package, the NOAO PTOOLS package, and the addon TABLES package can perform operations on the output photometry files produced by QPHOT and PHOT, including computing and displaying aperture dependent magnitude differences. The MKAPFILE and APFILE curve-of-growth analysis tasks in the PHOTCAL package also be used to examine and analyze the magnitude differences as a function of measuring aperture.

8.4. Computing the Standard Star Solutions

Computing the standard star solutions involves: creating a standard star catalog, creating a standard star observations file, defining the algebraic form of the transformation equations, and solving the transformation equations.

In IRAF creating a standard star catalog from scratch requires typing or reading the standard star name and photometric indices into a text file. The Landolt UBVRI standards (Landolt 1983, 1992), the Elias JHKL standards (Elias et. al 1982), and UKIRT JHK standards (Casali and Hawarden 1992) already exist on disk in a form that can be accessed by the photometric calibration code, and new catalogs can easily be added to the standard database.

The standard star observations file is constructed by combining the observations of each field in each filter. In IRAF this is done by creating an image sets definition file that assigns a name to each field and specifies which images belong to which field. The image sets definition file, and a user specified filter list and position matching tolerance are used to match up the calibration star observations. The fields in each filter may be offset from each other, there may be multiple standard stars per field, and there may be several observations of a single field. For the observation matching step to work successfully, it is critical that the correct exposure times, filter ids, airmass, and time of observation values were correctly read from the image headers and written to the photometry files. Mistakes can still be corrected and missing data supplied at this point, but the correction process is tedious and error prone.

For CCD data the transformation equations are usually defined in the "CCD sense", i.e. the photometric indices are on the right side of the equation, and the instrumental magnitudes are on the left. The equation solutions are inverted to solve for the photometric indices of the program stars. Discussions of the rationale behind this can be found in Stetson (1989) and Massey and Davis (1992). However in some cases the usual photoelectric formulation may do as well or better (Walker 1989). IRAF can formulate and solve for the equations using either convention.

Achieving good fits to the transformation equations requires that the standard stars have a reasonable spread in instrumental magnitude, color, airmass, and time of observation, one that brackets those of the program stars, if terms involving these quantities are to be fit. In some case users can use known color terms and fit only the zero point and extinction terms, or use mean extinction values, but it is a good idea not to rely on these quantities if doing precision work. After solving the transformation equations, users should remove deviant points from and refit the solution, and carefully check that there are no trends in the residuals with magnitude, color, airmass, or time of observation. Experience has shown that the precision and efficiency of CCD photometry should equal or exceed that of conventional photoelectric techniques for all but the brightest stars observed at high frequency (Walker 1989).

However users may find that the residuals in their transformation equation solutions are significantly larger than can be accounted for by the quoted errors in the catalogs and the formal errors in the photometry. At KPNO experience with the Landolt BVRI standards has shown that the fit residuals in each filter are typically 0.01-0.015 magnitudes with somewhat higher residuals in the B band compared to VRI. The probable cause of this phenomenon is a combination of the catalog errors being underestimated and mismatchs between the Landolt and KNPO CCD filter systems.

The IRAF tools used to do the photometric calibrations are located in the PHOTCAL package and are described in Massey and Davis (1992). Tools for computing relative shifts between images can be found in the IMMATCH package, see in particular the XREGISTER task. General purpose tasks for manipulating text files can be found in the addon ST TABLES package.

8.5. Finding Program Stars in Crowded Fields

In crowded program fields there are too many stars to identify by eye with the image display and image cursor, so some kind of automatic detection algorithm is required. Most such algorithms require knowledge of the shape and width of the stellar PSF width and a detection threshold.

IRAF uses the DAOFIND task in the DAOPHOT package to generate the star lists in crowded fields. DAOFIND searches the image for Gaussian intensity peaks whose FWHM is equal to a user supplied value and whose peak values are above some user supplied threshold above background. DAOFIND computes accurate pixel coordinates and shape parameters for all the detected objects, and outputs a final list of sources that are stellar or close to stellar. Details about the DAOFIND algorithm can be found in Stetson (1987), and a discussion of its IRAF implementation in Davis (1994).

8.6. Aperture Photometry of the Program Stars

If the program stars are uncrowded or moderately crowded they can be accurately measured using the standard aperture photometry techniques described in section 8.3. Even if the program fields are sufficiently crowded so that PSF fitting is required, aperture photometry is still required to provide initial magnitude and sky estimates to the PSF fitting code and to set the instrumental magnitude scale. In both case the photometry radius should be small, in order to maximize signal-to-noise ratios for the fainter program stars, and to minimize the effects of crowding. The optimal aperture radius varies somewhat with intensity and the shape of the PSF, but in many cases a value ~ 1 FWHM for typical ground-based data will yield good results. The program star instrumental magnitudes must eventually be transformed to the standard star instrumental magnitude scale by computing and applying an aperture correction using standard magnitude difference or curve-of-growth techniques (Howell 1989, Stetson 1992c).

Some care should be taken in estimating sky values for stars in crowded fields. In very crowded fields it may be necessary to position the sky annulus closer to the star than is normally done for isolated or moderately crowded stars to avoid severe contamination due to crowding. Where the crowding is especially severe or there are large background variations due for example to nebulosity or an underlying galaxy, better sky estimates may be obtained by changing the default IRAF / DAOPHOT "mode" sky algorithm to "median" or "centroid".

The standard fractional pixel aperture photometry algorithm used in IRAF is based on one originally developed for MPC, the KPNO Mountain Photometry Code, and later adopted by the POORMAN, DAOPHOT, and IRAF APPHOT and DAOPHOT photometry codes. It approximates a circle by an irregularly shaped polygon and produces good results for most choices of aperture radii. However aperture radii $<= 1.0$ should be avoided. The effective area of the fractional pixel apertures does depend on exactly where the center of a star falls on a pixel, an affect which can introduce scatter into the photometry. This effect should be negligible for radii $>= 2.0$ pixels. Users interested in doing aperture photometry through very small apertures or aperture photometry of undersampled data in moderately crowded fields should check out the addon package CCDCAP developed by Ken Mighell (1998).

8.7. PSF Fitting Photometry of Program Stars

If the program fields are so crowded so that the wings of neighboring stars can significantly affect the aperture photometry of the program stars, then PSF fitting techniques are required. There are many PSF fitting codes in the literature. IRAF implements the DAOPHOT II algorithms which are described in Stetson (1987), Stetson, Davis, and Crabtree (1989), and Stetson (1992a) in the form of an IRAF / DAOPHOT package (Davis 1994).

PSF fitting techniques are sometimes useful for measuring stars in uncrowded fields because: the optimal weighting scheme employed in PSF fitting codes can improve the signal-to-noise at the faintest magnitudes, accurate astrometry is a byproduct of the the fit, and the model fits permit the identification and rejection of bad data, the computation of goodness of fit statistics, and the option of easily subtracting off the star and examining the residual images.

8.8. Computing the PSF Model

Computing the PSF model is the most critical step in doing crowded field photometry. By default DAOPHOT uses a two component model of the PSF, an analytic component which should be chosen to match the data as closely as possible, and a look-up table of residuals subsampled by a factor of 2 with respect to the image. Options also exist to fit a purely analytic PSF model, which may minimize interpolation errors for undersampled data, or to compute additional look-up tables which describe linear or quadratic spatial variability of the PSF model. The code computes the PSF model by fitting the cores of selected bright isolated stars to the analytic function and modeling the residual intensity in the core and wings with the look-up table. It is critical that the PSF stars be bright unsaturated stars with no close neighbors or underlying detector blemishes. If a variable PSF model is used it is absolutely critical that the PSF stars be well distributed over the image. In very crowded fields fitting the PSF may be an iterative process of successively removing PSF star neighbors and recomputing the PSF on the neighbor subtracted image. Descriptions of how to model the PSF in crowded fields can be found in Stetson (1992b), Massey and Davis (1992), and Davis (1994).

8.9. Fitting the PSF to the Program Stars

DAOPHOT computes new positions and magnitudes for all the program stars in the input list by grouping the stars and fitting the stars in each group simultaneously using standard non-linear least squares techniques. The composition of the groups may change with each iteration as some stars are successfully fit and subtracted from the image. During the fitting process stars may be rejected because they are too faint, do not contain enough good data, or have merged with another star. At the conclusion of the fits chi and sharpness values are computed for each fitted star and the fitted star is subtracted from the input image. Assessing the quality of the fits is done by examining the final subtracted image and plotting the fitted chi values versus magnitude and position. Subtractions should be clean with residuals the size predicted by the noise model, and the fitted chi values should scatter around 1 with no obvious trends with magnitude or position in the image.

By default DAOPHOT does not refit the sky values in the PSF fitting step but instead uses the values computed in the initial aperture photometry step described in section 8.6. However experience suggests that in very crowded regions, refitting the sky can significantly improve the ability of the code to successfully fit the faintest stars as well as improve the quality of the fits (Parker 1991), and this option is available in DAOPHOT. For more detailed discussions of the problems of sky fitting in crowded regions and alternate techniques for dealing with variable background see Stetson and Harris (1988), Parker (1991), and Secker (1995).

Some experiments with artificial data also show that accurate initial guesses for the stellar positions can significantly improve the fits. For example if positional information from a high resolution ST image can be successfully transformed to the coordinate system of a lower resolution ground-based image, the ground-based image fits can be improved significantly.

8.10. Finding Hidden Program Stars

In most cases the subtracted image produced by the PSF fitting code will reveal the presence of hidden stars that were missed in the first iteration of DAOPHOT. Users must run the automatic star finding task on the subtracted image, get initial aperture photometry for the newly detected stars on the original unsubtacted image, merge the new aperture photometry list and the previous psf fitted list, and refit the stars using the merged list as input. In most case 2-3 iterations will suffice to pick up the majority of the stars of interest.

8.11. Calibrating the Program Stars

Computing calibrated magnitudes for program stars requires computing an accurate aperture correct for each program image as discussed in section 8.6. For images with non-variable PSFs this requires computing the magnitude difference between the measurements through the small crowded-field photometry aperture and the large standard star aperture and subtracting this value from the fitted crowded-field instrumental magnitudes. For images with variable PSFs the aperture correction computed this way is position dependent, and this method will not work. Instead the difference between the fitted magnitudes of the neighbor subtracted PSF stars and their magnitudes measured through the large standard star aperture must be computed and subtracted from the instrumental magnitudes. After applying the aperture correction the observations must be matched as described for the standard stars in section 8.4, and the transformations computed for the standard stars applied.

9. Data Modeling Tools

Data modeling tools can be used to test out reduction and analysis procedures before observing. This can be particularly useful if the observing instrument is new and may present new challenges to the software. For an excellent example of how to test software using artificially generated data see Holtzman's paper (1990) on testing the original version of DAOPHOT (Stetson 1987) with simulated WFPC data. Some of the enhancements suggested by those tests were incorporated into later versions of DAOPHOT (Stetson 1992a).

The IRAF ARTDATA package can be used to simulate realistic CCD calibration and star field data including realistic headers. Users intending to use this software for precision photometry tests should realize that the default setup for the package routines include trade-offs of accuracy, versus efficiency, and that the default parameters might require adjustment in favor of precision. Artificial images can also be created with the IMEXPR task.

DAOPHOT users can test the performance of the DAOPHOT routines on their particular data, by adding artificial stars of known position and magnitude to the data using the current PSF model, re-reducing the altered frame with DAOPHOT, and comparing the input and fitted positions and magnitudes of the artificial stars.

Ambitious users concerned about the precision of their photometric calibrations can try to model the photometric transformation they observe using data about their telescope, CCD, filters, exposure times, stars being measured, etc and the addon STSDAS package SYNPHOT.

10. Future Developments

A new GUI aperture photometry is currently under development. This task will eventually include support for: automatically generating and and saving object lists; annular or offset circular, elliptical, rectangular, and polygonal sky apertures; multiple circular, elliptical, rectangular, and or polygonal photometry apertures; simple shape and curve of growth analysis; and storing the results in a format compatible with existing IRAF photometry software. The GUI interface will support facilities for: selecting and displaying image and object list data; creating, viewing and editing object lists; setting, reviewing, editing and saving parameter sets; selecting objects to measure interactively or from the object list using the image display, image cursor, and region markers; displaying the results in both text table and graphical form; and interactive help text.

Acknowledgments. We are grateful to Philip Massey for suggesting that we use his 1998 paper as an example of how to use standard photometric techniques in IRAF environment to do science.

References

Abbott, T. M. C., et. al 1992, ApJ, 399, 680

Adams, M., et. al 1980, in Stellar Magnitudes from Digital Pictures, KPNO Internal Report

Armandroff, T., et. al, 1998, in NOAO CCD Mosaic Imager User Manual, http://www.noao.edu/kpno/mosaic/mosaic.html

Bushouse, H. 1995, The SYNPHOT User's Guide, http://ra.stsci.edu/documents/SyG_95/SG_1.html

Casali, M. & Hawarden, T. 1992, JCMT-UKIRT Newsletter, 4, 33

Chiu, L. -T. 1977, AJ, 82, 842

Davis, L. E. 1994, in A Reference Guide to the IRAF/DAOPHOT Package, http://iraf.noao.edu/docs/docmain.html

Davis, L. E., et. al 1996, in The CCDPhot Users Manual, http://www.noao.edu/kpno/manuals/ccdphot/

Elias, J. et. al 1982, AJ, 87, 1029

Goad, L. 1986, Proc. SPIE Int. Soc. Opt. Eng., Vol. 627, 688

Holtzman, J. 1990, PASP, 102, 806

Horne, K. 1988, in New Directions in Spectrophotometry, A. G. D. Philip, D. S. Hayes, S. J. Adelman, L. Davis Press, 145

Howell, S., 1989, PASP, 101, 616

Howell, S. B. 1992, in ASP Conf. Ser., Vol. 23, S. B. Howell, Astronomical CCD Observing and Reduction Techniques, 105

King, I., 1971, PASP, 83, 199

Koorneef, J., et. al 1985, in Highlights Of Astronomy, Vol. 7, J. -P. Swings, Dordrecht: Reidel, 833

Landolt, A. U. 1983, AJ, 88, 439

Landolt, A. U. 1992, AJ, 104, 340

Lasker B. M., et. al 1990, AJ, 99, 2019

Massey, P., et. al 1989, AJ, 97, 107

Massey, P. & Jacoby, G. H. 1992, in ASP Conf. Ser., Vol. 23, S. B. Howell, Astronomical CCD Observing and Reduction Techniques, 240

Massey, P. & Davis, L. E. 1992, in A User's Guide to Stellar CCD Photometry with IRAF, http://iraf.noao.edu/docs/docmain.html

Massey, P., et. al 1996, in An Observer's Guide to Taking Data with ICE, http://www.noao.edu/kpno/manuals/ice/ice.html

Massey, P., et. al 1997, in Direct Imaging Manual for Kitt Peak, http://www.noao.edu/kpno/manuals/dim/node1.html

Massey, P. 1997, in A User's Guide to CCD Reductions with IRAF, http://iraf.noao.edu/docs/docmain.html

Massey, P. 1998, ApJ, 501, 153

Mighell, K. 1998, see article this volume

Newberry, M. 1991, PASP, 103, 122

Parker, J. 1991, PASP, 103, 243

Scaller, S. 1992, in ASP Conf. Ser., Vol. 25, Astronomical Data Analysis and Systems I, D. M. Worrall, C. Biemesderfer, J. Barnes, 482

Secker, J. 1995, PASP, 107, 496

Shaw, H. & Horne, K. 1992, in ASP Conf. Ser., Vol. 25, D. M. Worrall, C. Biemesderfer, J. Barnes, Astronomical Data Analysis and Systems I, 482

Stetson, P. B. 1987, PASP, 99, 191

Stetson, P. B. & Harris, W. E. 1988, AJ, 96, 909

Stetson, P. B. 1989, in The Techniques of Least Squares and Stellar Photometry with CCDs, Proceedings of the vth Advanced School of Astrophysics, 51

Stetson, P. B., Davis, L. E., Crabtree, D. C. 1989, ASP Conf. Ser. 8, in CCDs in Astronomy, 289

Stetson, P. B. 1992a, ASP Conf. Ser., Vol. 25, D. M. Worrall, C. Biemesderfer, J. Barnes, Astronomical Data and Analysis and Systems I, 297

Stetson, P. B. 1992b, in User's Manual for DAOPHOT II

Stetson, P. B. 1992c, PASP, 102, 932

Stone, R. C. 1989, AJ, 97, 1227

Stover, R. J. & Allen, S. J. 1987 PASP, 99, 877

Tody, D. & Davis, L. E. 1992, in ASP Conf. Ser., Vol 25, Astronomical Data Analysis and Systems I, D. M. Worrall, C. Biemesderfer, J. Barnes, 484

Tody, D. & Valdes, F. 1998, to appear in Optical Astronomical Instrumentation, Proceedings of SPIE, Volume 3355

Valdes, F. 1996, in NOAO Image Data Structure Definitions,
    http:/iraf.noao.edu/projects/ccdmosaic/imagedef/imagedef.html#1
Walker, A. 1989, ASP Conf. Ser., Vol. 8, in CCDs in Astronomy, 319
Zarate, N. 1998, in IRAF Newsletter, Number 14,
    http://iraf.noao.edu/irafnews/apr98/irafnews.e.html

CCD Aperture Photometry

Kenneth J. Mighell

Kitt Peak National Observatory
National Optical Astronomy Observatories
P.O. Box 26732
Tucson, AZ 85726-6732

Abstract. This contribution describes how the basic techniques of CCD aperture photometry can be used to determine instrumental stellar magnitudes from calibrated and flat-fielded CCD observations of stars.

1. Introduction to CCD Aperture Photometry

The process of determining the brightness of a star can be surprisingly complex even with a technique as simple as aperture photometry. In principle, all one needs to do is add up all of the light near a star, subtract the contribution from the nearby sky background, and then convert the remaining number of photons to a stellar magnitude. This article describes how instrumental stellar magnitudes can be derived from calibrated and flat-fielded CCD observations of stellar fields using the techniques of CCD aperture photometry. Readers unfamiliar with the basic techniques of CCD stellar photometry are encouraged to read the review article of Da Costa (1992) and the references therein. The important process of transforming instrumental magnitudes to a standard system has already been extensively covered in the literature.

2. Determining the Centers of Stars

Most CCD aperture photometry programs require the user to give the position of the star on the CCD frame. If told to do so, most of these programs offer a choice of centering algorithms to determine the center of a star when provided with only a rough estimate. The review article of Stone (1989) compares the performance of five different digital centering algorithms under a wide range of atmospheric seeing and background-level conditions. If precise absolute or relatives positions from CCD observations are required, the reader should investigate the extensive literature dedicated to CCD astrometry.

3. Determining the Nearby Background

The background ("sky") flux associated with a star is generally determined by analyzing the distribution of intensities of nearby background pixels. In the case of circular aperture photometry, the background flux is typically determined by analyzing the pixels in an annulus beyond the stellar aperture. The inner radius of the sky annulus is typically several FWHM[1] distant from the center of the aperture in order to avoid the inclusion of contaminating light from the star itself. The width of the annulus is typically large enough so that the annulus contains a few hundred pixels. Many aperture photometry programs allow the user to set the background flux to be the modal value of the background intensity distribution. The mode of the background distribution is frequently estimated by using the following useful approximation,

$$\text{mode} \approx 3 \times \text{median} - 2 \times \text{mean} \quad (\text{for median} \lesssim \text{mean}), \tag{1}$$

of Pearson (1895). Unfortunately, this approximation is known to produce background estimates that are biased towards higher values (see, e.g., Newberry 1992). Fortunately, better methods for the estimation of the background are available. Modal estimates of the background distribution *after* the iterative rejection of outlier pixel intensities beyond 2.5–3.0 standard deviations of the mean generally produce reasonable results (Da Costa 1992). Many algorithms are available and most aperture photometry programs offer the user a choice of several methods to determine the background flux. For example, the popular APPHOT aperture photometry IRAF package offers the user the choice of 11 different ways the background can be estimated (Davis 1987). As always when using analysis software, the astronomer is strongly advised to carefully read the user documentation in order to understand which methodology will produce the best results for a given CCD observation.

4. Determining the Instrumental Stellar Magnitude

The fundamental task of a CCD aperture photometry program is to accurately measure all the electrons, S_A, that fall within an aperture placed on a CCD image. Unless the background flux, B, is exactly zero electrons per pixel, one can never directly measure the number of electrons, $S_\star$, from a star within the aperture. One measures, instead, the quantity $S_A = S_\star + N_A B$, where N_A is the aperture of the aperture (in pixels). The instrumental magnitude of an aperture measurement of a CCD observation of a star can thus be defined as

$$\begin{aligned} m &\equiv -2.5 \log \left(\frac{S_\star}{1\,\text{e}^-} \right) \\ &= -2.5 \log \left(\frac{S_A - N_A B}{1\,\text{e}^-} \right) \end{aligned} \tag{2}$$

where the latter form is in terms of observable quantities. By this definition, a stellar intensity of one electron would have an instrumental magnitude of zero. The estimated error of the instrumental magnitude is approximately

$$\Delta m \approx 1.0857 \left[\frac{\Delta S_\star}{S_\star} \right] \tag{3}$$

---

[1] The Full-Width-at-Half-Maximum of a critically-sampled Gaussian distribution is ∼2.36 pixels.

where $\Delta S_\star$ is the measurement error (in electrons) of the estimated number of electrons from the star within the aperture.

It is clearly important to determine the observable quantities S_A, B, and N_A as accurately as possible. Let us consider the aperture area, N_A, first. If a_j is the area of the jth aperture pixel, then the total area of the aperture, N_A, is simply

$$N_A = \sum_{j=1}^{A} a_j , \qquad (4)$$

where A indicates that the summation includes all pixels or partial pixels within the aperture. The area of the jth aperture pixel will be exactly one pixel only when the pixel is completely within the aperture, otherwise only a fraction of the pixel lies within the aperture and the value of a_j is between zero and one ($0 < a_j < 1$ pixel). The area of a *circular* aperture is $A = \pi r^2$ pixels where the radius of the aperture is r pixels. Although it is not difficult to exactly determine the partial pixel areas with square pixels and circular apertures, several standard CCD aperture photometry programs only approximate the partial pixel areas. For example, the popular PHOT task of the APPHOT package approximates the circular aperture by an irregular polygon (Davis 1987). While this approximation is generally fine with large apertures, it can sometimes produce large systematic measurement errors for small aperture radii.

Let us now consider the sum of all electrons within the aperture (S_A). If z_j is the intensity (in electrons) of the jth aperture pixel, then one simple approximation of S_A is

$$S_A = \sum_{j=1}^{A} a_j z_j . \qquad (5)$$

This approximation linearly weights the pixel intensity with the area of the pixel within the aperture. The PHOT task uses a linear pixel weighting algorithm that is very similar to this approximation. The use of Equation (5) implicitly assumes that the Point Spread Function (PSF) is nearly flat at the edge of the aperture. The PSF is generally nearly flat only at large distances from the center of a star where the encircled energy function is nearly equal to one (i.e. 100%). The use of Equation (5) is thus generally appropriate for large aperture radii. Wherever the PSF is a rapidly changing function of radius (e.g., at small radii of seeing-optimized CCD observations) the use of Equation (5) is likely to produce large systematic measurement errors.

The precision of any given CCD aperture photometry algorithm can be determined by analyzing how accurately a two-dimensional Gaussian distribution can be measured. A Gaussian is a good model PSF of a ground-based CCD observation since the central core of a ground-based stellar profile is approximately Gaussian (King 1971). Let us assume that the PSF is a two-dimensional Gaussian function

$$\phi(r;\sigma) = \frac{1}{2\pi\sigma^2} e^{-\frac{1}{2}\left(\frac{r}{\sigma}\right)^2} \qquad (6)$$

where the radius, r, and the standard deviation, σ, have pixel units. The encircled energy function of this PSF is then

$$\mathrm{EE}(r;\sigma) = \int_0^r \int_0^{2\pi} \phi(\rho;\sigma)\, \rho\, d\rho\, d\theta = 1 - e^{-\frac{1}{2}\left(\frac{r}{\sigma}\right)^2}. \qquad (7)$$

The systematic measurement error associated with the aperture radius is

$$\frac{\partial\, \mathrm{EE}(r;\sigma)}{\partial\, r} = \frac{r}{\sigma^2}\, e^{-\frac{1}{2}\left(\frac{r}{\sigma}\right)^2}\, dr \qquad (8)$$

which will be small for large radii or large standard deviations. Numerical experiments which have shown that the systematic measurement error, dr, is ~ 0.25 pixel for critically-sampled Gaussians ($\sigma \equiv 1$) when Equation (5) is used (Mighell & Rich 1995). This translates to a photometric errors of 0.122, 0.068, 0.027, 0.008 mag for radii of 1.5, 2, 2.5, and 3 pixels, respectively.

One way of improving the dr systematic measurement error is to split each pixel into subpixels by using a bilinear pixel interpolation algorithm. One such algorithm is to use the two-dimensional analog of the sinc function: $J_1(\pi r)/2r$. This function, like many others, has the unfortunate effect of degrading the original image by spreading photons (electrons) beyond the original pixel. Mighell has created the QUADPX bilinear pixel interpolation algorithm (see Fig. 18 of Mighell & Rich 1995) which splits a pixel into 4 subpixels whose sum is always equal to that of the original pixel. The QUADPX algorithm has one free parameter, the gradient hardness parameter, H, which ranges from 0.0 to 1.0. By definition, using the QUADPX algorithm with H≡0.0 is equivalent to using Equation (5). The appropriate H value generally depends on the sampling of the PSF and the size of the aperture. In general, one finds with small apertures that H≈1 gives the optimal performance (minimal systematic measurement error) for undersampled PSFs while a value of H≈0 is optimal when analyzing heavily oversampled PSFs. Numerical experiments have shown that using H≡0.8 with critically-sampled Gaussians gives a systematic measurement error of $dr \approx 0.04$ pixels which translates to photometric errors of 0.019, 0.011, 0.004, 0.001 mag for radii of 1.5, 2, 2.5, and 3 pixels, respectively (Mighell & Rich 1995). The systematic measurement error of critically-sampled Gaussians can be reduced by a factor of six through the simple expedient of using H≡0.8 instead of H≡0.0. The QUADPX algorithm has been implemented in the CCDCAP[2] digital circular aperture photometry code developed by Mighell (1997, and references therein) to analyze *HST* WFPC2 observations of extragalactic stellar populations.

5. Optimal Aperture Size and Aperture Corrections

The best (smallest) stellar photometric errors (i.e. the largest signal-to-noise ratios) are generally obtained with relatively small apertures (see, e.g., Fig. 6 of Howell 1989). Analysis of theoretical CCD signal-to-noise-ratio equations (see,

---

[2] IRAF implementations of CCDCAP are now available over the Wide World Web at the following site: http://www.noao.edu/noao/staff/mighell/ccdcap

e.g., Newberry 1991, Howell 1992, Merline & Howell 1995, Howell et al. 1996, and references therein) shows that large apertures can have large photometric errors when the the total number of stellar photons in the aperture becomes comparable with the total number of background photons in the aperture. Furthermore, a measurement error for the background flux as small as just 1 electron per pixel can by itself produce large photometric uncertainties at large aperture radii. Small apertures, however, can be *too* small when they allow such a small fraction of the star light to be found within the aperture that the photometric error becomes dominated by small-number (a.k.a. counting or Poisson) statistics (i.e. when little or no signal is measured).

One can easily show that the optimum signal-to-noise ratio for a Gaussian Point Spread Function, $\phi(r;\sigma)$, is obtained for a circular aperture radius of $r_0 \approx 1.6\sigma$ (i.e. $r_0 \approx 0.68\,\text{FWHM}$) which contains about 72% of the total light from the star according to Equation (7). Pritchet & Kline (1981) note that the signal-to-noise ratio is fairly insensitive to radius near the "optimal" radius value $\sim 1.6\,\sigma$ for a Gaussian PSF; deviations from the optimal radius by as much as ±50% make little difference. Since centering errors will be more critical for smaller apertures than larger apertures, it is only prudent to err on the larger side by using apertures with radii which are larger than $\sim 0.68\,\text{FWHM}$.

An aperture radius of $r \approx \text{FWHM}$ makes an excellent practical compromise between concerns about systematic centering errors and diminishing signal-to-noise ratios typically obtained with larger aperture radii. By analyzing the CCD signal-to-noise ratio equations, one can show that brighter stars will have larger optimal aperture sizes than do fainter stars. If one must use only one aperture size, then it is clearly advantageous to chose a global aperture size which produces the smallest photometric errors for the faintest stars (i.e. use $r \approx \text{FWHM}$).

Small apertures frequently do not contain all the flux from a star. The amount of the missing star light can found by determining the appropriate aperture correction by measuring nearby bright isolated stars. Howell (1989) and Stetson (1990) describe the process how aperture corrections can be accurately determined using the aperture growth-curve method.

Acknowledgments. KJM was supported by a grant from the National Aeronautics and Space Administration (NASA), Order No. S-67046-F, which was awarded by the Long-Term Space Astrophysics Program (NRA 95-OSS-16). This research has made use of NASA's Astrophysics Data System Abstract Service and the NASA/IPAC Extragalactic Database (NED) which is operated by the Jet Propulsion Laboratory at California Institute of Technology, under contract with NASA.

References

Da Costa, G. S. 1992, in Astronomical CCD Observing and Reduction Techniques, ASP Conference Series, 23, 90

Davis, L. 1987, "Specifications for the Aperture Photometry Package", (internal NOAO publication) [ftp://iraf.noao.edu/iraf/docs/apspec.ps.Z]

Howell, S. B. 1989, PASP, 101, 616

Howell, S. B. 1992, in Astronomical CCD Observing and Reduction Techniques, ASP Conference Series, 23, 105

Howell, S. B., Koehn, B., Bowell, E., & Hoffman, M. 1996, AJ, 112, 1302

King, I. R. 1971, PASP, 83, 199

Merline, W. J., & Howell, S. B. 1995, *Experimental Astronomy*, 6, 163

Mighell, K. J., & Rich, R. M. 1995, AJ, 110, 1649

Mighell, K. J. 1997, AJ, 114, 1458

Newberry, M. V. 1991, PASP, 103, 122

Newberry, M. V. 1992, in Astronomical Data Analysis Software and Systems I, ASP Conference Series, 25, 307

Pearson, K. 1895, *Philosophical Transactions*, 186, 343

Pritchet, C., & Kline, M. I. 1981, AJ, 86, 1859

Stetson, P. B. 1990, PASP, 102, 932

Stone, R. C. 1989, AJ, 97, 1227

Point-Spread Function Fitting Photometry

J. N. Heasley

Institute for Astronomy, University of Hawaii, 2680 Woodlawn Drive, Honolulu, HI 96822

Abstract. Point-spread function fitting (PSF) photometry has proven to be a valuable tool for determining the magnitudes of stars in crowded fields. The basics of PSF fitting photometry are reviewed with an eye toward new and potential users, emphasizing the practices common to most of the programs that have been developed for this purpose.

1. Introduction

One of the most productive interactions between observations and theory in astronomy has been in the field of stellar evolution. High-precision photometry of large numbers of stars in clusters has provided critical tests for models of stellar structure and evolution. The introduction of charge-coupled devices (CCDs) in the late 1970s provided observational astronomers with high-quantum efficiency, linear area detectors that were quickly turned toward problems requiring precision photometry of large samples of stars.

As discussed by Leach during this workshop, CCD technology has advanced greatly from the early pioneering efforts. Detectors are now pushing the envelopes in terms of high-quantum efficiency and low read-noise. The physical dimension of these sensors has grown significantly, and in many instances these devices are being incorporated into arrays that allow even large fields to be observed at the telescope. It is now possible for an observer to generate over a billion pixels of photometric quality data *per night*. Advances in detector technology are also making large-format imaging devices in the infrared a reality.

Fortunately, the computer power needed to process the volume of photometric data produced by modern CCDs has grown in parallel with the ability of the devices to generate the data. An astronomer now can have a desktop computer with the numerical processing power of what only 10–12 years ago was considered a "supercomputer."

Equally important to the power of one's computer hardware are the algorithms one uses to process the data. As observers have pushed their observations to fainter and more compact objects, simple photometric approaches (e.g., aperture photometry) were unable to yield the photometric precision desired, and more elaborate approaches such as point-spread function (PSF) fitting were developed. These procedures have become quite sophisticated over many years of testing and application to astronomical data by many individuals and groups.

While I was originally asked to talk about the PSF software that Dr. Kenneth Janes (Boston University) and I have developed (Janes & Heasley 1993), I decided to concentrate on a brief review of the basics of PSF stellar photometry that are common to all of the software packages developed for this purpose. Detailed discussions of many of the topics touched upon here are covered by Stetson (1987), Massey & Davis (1992), and in the list of selected references provided by Davis in her talk at this workshop.

2. PSF Photometry

2.1. Basic Assumptions

In performing PSF photometry, we make two basic assumptions:

1. All point sources imaged by the atmosphere–telescope–detector can be represented by a point-spread function. If this function is *not* spatially invariant, we assume that it is knowable and can be modeled.
2. One's imaging system is linear in its response to input radiation. Observers must understand the limitations of their detector systems.

While solid state imaging devices are a vast improvement over the photographic process, these detectors do have an upper limit on their linearity. Further, their dynamic range is not infinite, and care must be taken in developing one's observing program to ensure the total range of brightness one wants to measure is covered with adequate signal-to-noise.

With these assumptions in mind, the basics of PSF photometry become quite straightforward. Given that the shape of a star on the focal plane is known, its amplitude will scale linearly with the brightness of the star forming the image. Thus, for any stellar image, we must "fit" the PSF template to the observed image, allowing for the fact that the stellar image sits on top of some sky background. In principle the process is simple: We must identify the stars within our CCD image, determine the sky level under each star, use "isolated" stars to derive the PSF template, and then fit this template to all the stars in the frame. In practice, however, this procedure must often be repeated several times as one develops an improved model of the PSF and sky levels.

2.2. Finding Stars

Numerous algorithms have been developed for finding stars in digital images. Many of these use a cross correlation approach wherein the PSF template (or a reasonable facsimile thereof) is used to detect stars. As the PSF is passed over the image, peaks will occur in the cross correlation at the location of stars in the image. The amplitude of the peak above the level one gets for "pure" sky will depend on the brightness of the star and the noise level of the sky. Some programs take the initial star finding further (e.g., Stetson 1987). They include measures of the "sharpness" of the source in order to determine whether the object found by the process is truly stellar or is a faint galaxy. For an initial search of an image, if the PSF template has yet to be created, a two-dimensional Gaussian with the appropriate full width at half-maximum (FWHM) will usually yield good results.

Even the best star-finding algorithms will sometimes fail to find faint stellar companions located next to bright stars. These stars can usually be found on a second pass through the image after the stars found in the first detection pass have been measured by the PSF fitting and then subtracted from the image. By running one's finding routine on this residual image, the stars missed on the previous pass will usually be located. Usually, two or three cycles of finding and fitting will detect virtually all the stars in a frame, with the exception of some pathological cases, e.g., extremely close blends of stars of nearly identical magnitude. These stars can usually be handled interactively by the astronomer with an image display to pinpoint the initial estimates of their locations.

2.3. Finding the Sky Level

Determining the sky level under a stellar source has received considerable attention in the literature. This problem is especially important for faint stars where small errors in the sky level can translate into large uncertainties in the derived instrumental magnitude. Many of the PSF photometry programs determine the sky brightness independently of the stellar magnitude by examining the pixels in an annulus centered on the star and using some statistic of the distribution of sky brightness in this region to characterize the sky under the star in question. Eaton (1989) found that for automatic photometry, the mode and median of the distribution (with rejection) provided reliable estimators of the sky level. This approach is fast and can be accomplished once for each star.

There are circumstances when this simple approach to deriving the local sky level for a particular star might fail. One clear example that is likely to cause problems arises when there are steep gradients in the background level, as one might encounter near the center of a globular cluster due to the light from unresolved stars. In such cases, an approach described by Stetson & Harris (1988) can often improve the results of the PSF fitting. After deriving initial estimates of the sky and fitting the stars in the frame, all the measured stars are subtracted from the image and the residual frame smoothed with a box median filter (of dimension comparable to the size of the PSF), and this smooth sky frame can be subtracted from the original frame. A constant sky value is then added to the frame to preserve the appropriate brightness level of the sky.

In the most recent implementation of our photometry program, Janes and I use a variation on this approach. We use the residual frame (after the stars that have been fitted are subtracted) to perform new estimates for the sky levels using a circular sky aperture. In this form, the sky distribution is "cleaner" because contaminating stars in the sky annulus have been removed before the distribution is examined to find the sky level.

Aurière & Coupinot (1989) have experimented with a variant of unsharp masking to measure stars in crowded fields with steep backgrounds.

2.4. Building the PSF

Determining the PSF for a particular frame is the most difficult, time consuming, and *important* part of the PSF photometry method. This process is often iterative in that successively improved versions of the PSF will be generated and applied to the frame.

To generate a good PSF, we require

- stars that are well isolated from their neighbors.

- stars with high signal-to-noise.

In crowded fields, there may be many stars with adequate signal-to-noise, but they often have close companions that compromise their usefulness as stars for generating a PSF. In such instances, one begins with a crude estimate of the PSF (perhaps even based upon a single isolated star) and uses it to fit the stars in the frame and remove the close neighbors of the candidates one would like to use in generating a newer version of the stellar template. In our SPS program, Janes and I have automated this process in that stars that have already been fitted and are close to a PSF candidate star are removed from the frame before the the candidate is extracted and incorporated into the new PSF template.

The normalization of the PSF is arbitrary—the PSF describes the shape of the stellar profile, not its amplitude. The scaling of the PSF to the individual stellar profiles is what defines the instrumental magnitudes. Thus, the magnitude system for the PSF photometry is defined within an arbitrary constant. Most users define the PSF to some instrumental scale by adding a constant that makes the resulting instrumental magnitudes come out in some reasonable numeric range and then apply *aperture* corrections to determine the offset between that system and the instrumental system used to measure photometric standard stars (derived from aperture photometry). In the SPS code, Janes and I have chosen to normalize our PSF directly to the aperture photometry at the same radius at which the standard stars were measured. While this eliminates the need for explicit aperture corrections, one should keep in mind that an aperture normalization is still done.

2.5. The Fitting Process

The actual fitting of the PSF template to an individual star identified by the finding procedure involves determining three parameters: the center of the star image on the array and the amplitude by which the PSF must be scaled to give the best fit (in a least-squares sense) to the observed star. If the star is reasonably isolated, one can determine the centering of the star by some a priori means (e.g., by looking at the marginal distribution). In such an instance the fitting problem is simple a linear least-squares problem. If, however, we choose to solve for the location of the star's center, the fitting becomes a nonlinear problem.

In general, one uses only the pixels located near the center of the star image, where the signal-to-noise is greatest, in the actual fitting process. Using the highest signal-to-noise pixels in the stellar profile is not unique to PSF fitting, and its importance in aperture photometry has been discussed by Howell (1989). Typically, the fitting radius is on the order of the FWHM of the stellar profile. It is also possible to formulate the least-squares problem to include the determination of the local sky as part of the solution. To do so, however, one must generally increase the region around the star included in the fit to include more sky pixels. These, of course, add little information to the solution for the amplitude parameter.

In DAOPHOT one can solve for the sky level under an individual star as part of the fitting process. In the case of crowded groups of stars, DAOPHOT can also determine a mean sky level appropriate for the entire group as part of the solution for the amplitudes.

3. Implementations

There have been many programs developed for performing PSF photometry, especially in crowded fields. This has resulted in the sharing of many working techniques among those developing such programs. Indeed, many of the programs listed here have developed independent implementations of the same basic algorithms.

In his paper describing the well-known DAOPHOT program, Stetson (1987) reviewed the early history of PSF photometry programs, including ROMAPHOT (Buonanno et al. 1983) and STARMAN (Penny & Dickens 1986). All three programs have been used widely in the astronomical community. More recent additions in the field include DoPHOT (Mateo & Schechter 1989), INVENTORY (Kruszewski 1989), CAPELLA (Llebaria et al. 1989), MOMF (Kjeldsen & Frandsen 1992), and SPS (Janes & Heasley 1993). Davis and her collaborators at NOAO have developed an independent implementation of the DAOPHOT algorithms in the context of the IRAF software environment (see, e.g., the references cited in her presentation at this workshop). There are almost certainly other programs for performing PSF photometry that I'm not aware of.

Virtually all of these codes are constantly under development as the authors improve them for their own research. The references above point to the initial description in the literature, not the most recent ones. In particular, DAOPHOT (both the original and IRAF versions), DoPHOT, and SPS all were originally limited to spatially invariant PSFs, but all have since been generalized to relax that requirement.

4. Complications

4.1. Severe Crowding

While PSF fitting programs are designed to work in "crowded" fields, observers are always pushing the limits and finding scientifically interesting targets that often stress the limits of a program's ability to resolve the stars and measure their magnitudes. In principle, the simultaneous solution of magnitudes for a group of overlapping stars is straightforward—for N stars we must determine $3N$ parameters and perhaps a common sky level beneath them. The practical problem is how big a grouping of stars does one solve simultaneously. If one is working on the ith star in the list of objects, intuitively one knows that the group should be large enough to include those stars that will "significantly" affect the target star, but not so large as to include those so far away that they are a minor or insignificant influence on it. Just how such groups fall on the detector depends on the region being observed, and it is unclear whether there is an optimal strategy for solving this problem—indeed, one could always invent pathological cases to thwart almost any approach!

How one solves the multiple star problem numerically is another problem that has been approached in different ways. DAOPHOT (Stetson 1987) solves the simultaneous equations for the group directly. In SPS (Janes & Heasley 1993), we approach the process iteratively, beginning with the brightest star in the group and working fainter in magnitude until all the stars have been measured, and if necessary, repeating the process until all instrumental magnitudes for the group stabilize within some set tolerance. In practice, for practical reasons one limits the maximum size of the group. Both approaches appear to work well on real data.

4.2. Spatial Variation of the PSF

A second complication in PSF photometry arises when the PSF varies spatially over the image. In the early generations of CCDs, the physical dimensions of sensors were so small that only a small portion of the focal plane was imaged. This is no longer the case, especially as one moves to arrays of multiple CCDs. We now must deal with the aberrations of the optical system in defining the PSF.

Many of the programs for PSF photometry (e.g., DAOPHOT, SPS, MOMF) implement the stellar template by using a Gaussian core (or some other analytic function) to remove the high spatial frequency component of the PSF leaving a table of more or less smoothly varying residuals. To account for the spatial variation of the profile, the most recent implementation of DAOPHOT (and the IRAF version of the algorithms) adopts a quadratic model for the residual corrections (Stetson 1991). The correction to the Gaussian core (i.e., the residual) is modeled as

$$\begin{aligned} C(i,j) &= C_0(i,j) + (X - X_0)C_1(i,j) + (Y - Y_0)C_2(i,j) + \\ &\quad (X - X_0)^2 C_3(i,j) + (Y - Y_0)^2 C_4(i,j) + \\ &\quad (X - X_0)(Y - Y_0)C_5(i,j) \end{aligned}$$

In SPS we have adopted this same model. There is certainly nothing physical about this model, and it is adopted primarily for computational simplicity. In principle, if one knew the aberrations of the optical system in advance, it would be possible to develop a more physical representation of the PSF to match the telescope-detector system. Of course, such a version would not be "general." In practice, with SPS we find the quadratic model does a good job.

If the PSF is spatially variable, we must have stars that sample that variation around the frame, or at the very least, those portions of the frame in which we wish to measure stars. This can become important depending upon how the variation of the PSF is described. For the quadratic model described above, if only one region of the frame were used to define the PSF residuals, one might get unexpected (and unrealistic) behavior as the coefficients extrapolate to other parts of the frame!

5. Photometry Cookbook

The inputs required for most PSF photometry programs are quite similar. One generally needs to specify information about the CCD sensor (e.g., linear full well, gain, read-noise), an estimate of the stellar FWHM, the size of the PSF array, the size of the fitting radius, aperture photometry radii, the sky annulus region, etc. The size of the fitting radius will be less than the size of the PSF array as one generally uses only the pixels near the center of the PSF (i.e., those with high signal-to-noise) for determining the instrumental magnitudes. The

PSF, however, should be defined over a larger area so that the stars that have been fit can be completely subtracted from the image. An excellent discussion of how one determines the appropriate values for many of these parameters is given by Massey & Davis (1992). While their exposition is directed toward using IRAF/DAOPHOT, the general principles apply to all PSF photometry programs.

While each set of CCD data is likely to have its own quirks, the following sequence of steps usually gives a good first cut at obtaining PSF photometry for an image.

1. Make an initial pass through the image with the star-finding procedure.
2. Determine the sky level for each star in the list and perform aperture photometry on all objects.
3. Generate an initial estimate of the PSF using the stars with the "best" aperture photometry (e.g., estimated errors in the instrumental magnitudes of less than 0.01 mag).
4. Use the preliminary PSF to fit all the stars to obtain more accurate instrumental magnitudes.
5. Subtract the stars that have been measured from the frame and run the star finding algorithm on the residual frame to locate stars missed during the initial pass.
6. Fit the newly discovered stars and any stars from the original list whose magnitudes change because of companions previously undiscovered.
7. Subtract all the stars from the frame and redetermine the sky levels for each star using the residual image.
8. Refine the estimate of the PSF. More PSF candidate stars should be available since one can now remove close companions and "isolate" stars suitable for defining the PSF.
9. Refit all the stars in the list.
10. Repeat the sky estimation and PSF refinements (steps 7–8) if needed.
11. Hand clean the image interactively using an image display if needed.

6. Variations on the Theme

An interesting variation on PSF fitting photometry has been developed by several groups for processing *Hubble Space Telescope (HST))* WFPC2 frames. A PSF fitting process is performed on the stars within a frame, perhaps even using a spatially invariant PSF. This PSF and the instrumental magnitudes are then used to remove neighboring stars around a star of interest so that one can perform simple aperture photometry on the target star. Thus, the PSF is only used to clean out neighboring stars, not to derive the instrumental magnitudes directly. Rubenstein (1997) used this approach with the original version of SPS (with spatially invariant PSF) to search for main-sequence binary stars in globular clusters.

Some other photometry-related issues discussed at this workshop also deserve some comment, particularly, how one deals with very bright stars. Several speakers have talked about using defocused images to spread out the light from the brightest stars so as to not saturate the CCD. Ultimately, this goes back to the finite full-well and limited dynamic range of a CCD. Certainly the approach using defocused images is a viable one and has been discussed by a number of authors (e.g., Kjeldsen & Frandsen 1992; Gilliland et al. 1993). Given the amount of effort astronomers have invested in minimizing seeing to produce ever sharper images, it seems rather wasteful to be forced to defocus frames containing bright stars. Indeed, in some instances, such as observations with the *HST*, the defocus option may simply not be practical. If one's CCD system has a shutter of adequate speed and precision, a viable alternative to defocusing is to image one's fields with exposures of different duration. Given the characteristics of one's CCD, the exposure times can be selected so as to not saturate the brightest stars in the field and to provide reasonable overlap in magnitudes between exposures.

7. Examples

In this section I will show several examples of PSF photometry that were processed using the SPS program. Figures 1 through 3 show different stages of PSF fitting for an image of NGC 103 obtained at the Kitt Peak 0.9 m telescope with a Tek 1024 × 1024 CCD. The original image and 2 stages of star-fitting and subtraction are shown. The last image in the sequence resulted from human interactive cleaning of the third residual frame.

My second example uses the globular cluster M92. The data come from the HRCam (McClure et al. 1991) at the Canada-France-Hawaii Telescope (CFHT) with a 1024 × 1024 SAIC CCD and the Kitt Peak 0.9 m telescope using a Tek 2048 × 2048 CCD. The HRCam images were used to measure the colors and magnitudes of main-sequence stars within the M92 video camera/CCD standard star region (Christian et al. 1985) while the KPNO 0.9 m images covered a large region around the cluster. The color-magnitude diagram shown in Figure 4 combines data on the brighter stars ($V < 17$) from the KPNO 0.9 m with deep images of the "video camera standard field" in M92 obtained with the HRCam at CFHT. The main sequence in this plot compares favorably with the M92 color-magnitude diagram of Stetson & Harris (1988) although the total integration times were shorter.

My final example uses the thick-disk globular cluster NGC 6637. Figures 5 and 6 show a V image of the cluster obtained with the CFHT and HRCam and a color-magnitude diagram using data from the HRCam observing run. Figure 7 shows the region in NGC 6637 covered by two *HST* PC frames that were obtained as part of our photometry program on clusters located near the Galactic center. This region overlaps with the CFHT frame in Figure 5. The color-magnitude diagrams from the *HST* observations are shown in Figure 8 (the frame further out from the cluster center) and in Figure 9 for those stars in the small overlap region of the *HST* pointings.

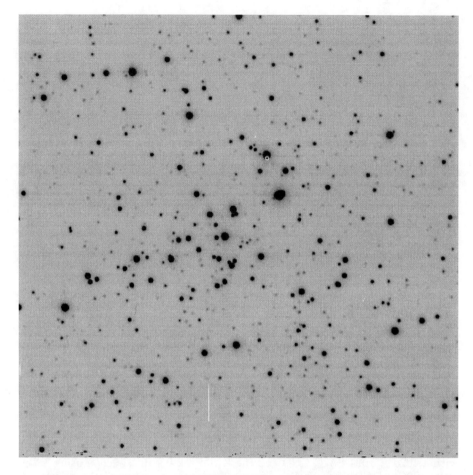

Figure 1.   CCD image of NGC 103 taken at the KPNO 0.9 m telescope.

8. Summary

PSF fitting photometry is the method of choice when performing stellar photometry in crowded fields. Those interested in exploiting the method have a variety of photometry packages available to them: the stand alone DAOPHOT, ROMAPHOT, DoPHOT, IRAF/DAOPHOT, and SPS, to name only of few of the programs that have been used widely in the astronomical community. While most of these programs share a common philosophy and approach, the details of how they implement various algorithms vary from program to program.

New users should understand the general principles involved with PSF fitting and the characteristics of their instrumentation and images rather than treating any of these programs as a "black box." As with any piece of software, the user should make sure that the answers it produces make sense.

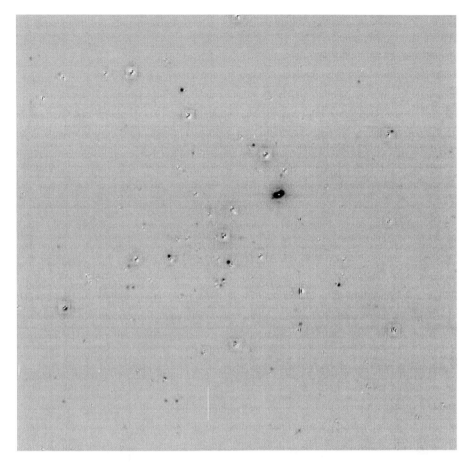

Figure 2. Same as Figure 1 after one pass of star-finding and fitting with SPS.

Acknowledgments. I would like to acknowledge many years of discussions on PSF photometry with my collaborator Dr. Kenneth Janes of Boston University.

Support for this work was provided by NASA through grant number GO-5366.04-93A from the Space Telescope Science Institute, which is operated by the Association of Universities for Research in Astronomy, Inc., under NASA contract NAS5-26555.

References

Aurière, M., & Coupinot, G. 1989, in Proceedings of the 1st ESO/ST-ECF Data Analysis Workshop, edited by P. J. Grosbøl, F. Murtagh, & R. H. Warmels (European Southern Observatory, Garching), p. 101

Figure 3. Same as Figure 2 after two passes of star-finding and fitting and final cleaning of the remaining stars by hand.

Buonanno, R., Buscema, G., Corsi, C. E., Iannicola, G., & Fusi Pecci, F. 1983, A&A, 51, 83.

Christian, C. A., Adams, M., Barnes, J. V., Butcher, H., Hayes, D. S., Mould, J. R., & Siegel 1985, PASP, 97, 363

Eaton, N. 1989, in Proceedings of the 1st ESO/ST-ECF Data Analysis Workshop, edited by P. J. Grosbøl, F. Murtagh, & R. H. Warmels (European Southern Observatory, Garching), p. 93

Gilliland, R. L., Brown, T. M., Kjeldsen, H., McCarthy, J. K., Peri, M. L., Belmonte, J. A., Vidla, I., Cram, L. E., Palmer, J., Frandsen, S., Parthasarathy, M., Petro, L., Schneider, H., Stetson, P. B., & Weiss, W. W. 1993, AJ, 106, 2441

Howell, S. B. 1989, PASP, 101, 616

Janes, K. A., & Heasley, J. N. 1993, PASP, 105, 527

Kjeldsen, H., & Frandsen, S. 1992, PASP, 104, 413.

Kruszewski, A. 1989, in Proceedings of the 1st ESO/ST-ECF Data Analysis Workshop, edited by P. J. Grosbøl, F. Murtagh, & R. H. Warmels (European Southern Observatory, Garching), p. 29

Llebaria, A., Perichaud, L., Leporati, L., & Debray, B. 1989, in Proceedings of the 1st ESO/ST-ECF Data Analysis Workshop, edited by P. J. Grosbøl, F. Murtagh, & R. H. Warmels (European Southern Observatory, Garching), p. 85

Massey, P., & Davis, L. 1992, "A User's Guide to Stellar CCD Photometry with IRAF"

Mateo, M. & Schechter, P. 1989, in Proceedings of the 1st ESO/ST-ECF Data Analysis Workshop, edited by P. J. Grosbøl, F. Murtagh & R. H. Warmels (European Southern Observatory, Garching), p. 69

McClure R. D., Arnaud, J., Fletcher, J. M., Nieto, J-L., & Racine, R. 1991, PASP, 103, 570

Penny, A. J., & Dickens, R. J. 1986, MNRAS, 220, 845

Rubenstein, E. 1997, Ph.D. thesis, Yale University

Stetson, P. 1987, AJ, 99, 191

Stetson, P. 1991, in Proceedings of the 3rd ESO/ST-ECF Data Analysis Workshop, edited by P. J. Grosbøl & R. H. Warmels (European Southern Observatory, Garching), p. 69

Stetson, P., & Harris, W. E. 1988, AJ, 96, 909

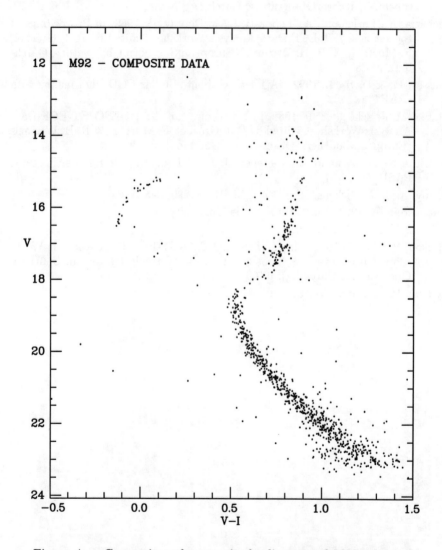

Figure 4. Composite color-magnitude diagram of M92 where the main-sequence stars have been measured on HRCam frames and the brighter evolved stars have been measured on frames from the KPNO 0.9 m telescope.

Point-Spread Function Fitting Photometry

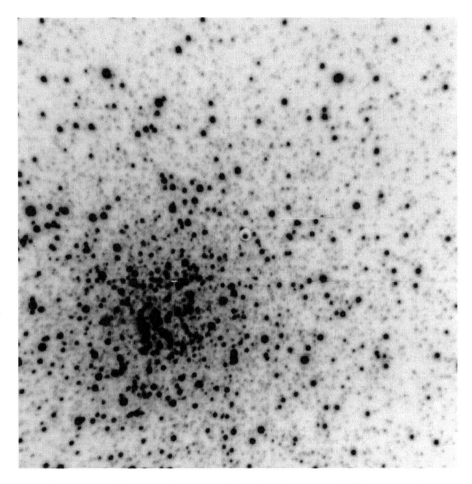

Figure 5. The central region of the globular cluster NGC 6637. This V band image was obtained with the HRCam at the CFHT.

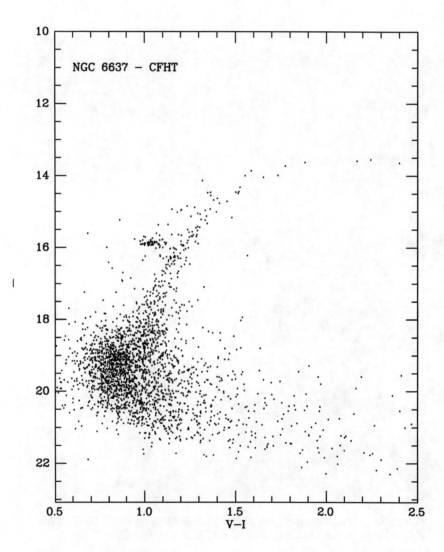

Figure 6. Color-magnitude diagram of NGC 6637 derived from HRCam observations.

Point-Spread Function Fitting Photometry 71

Figure 7. NGC 6637; HST PC frame.

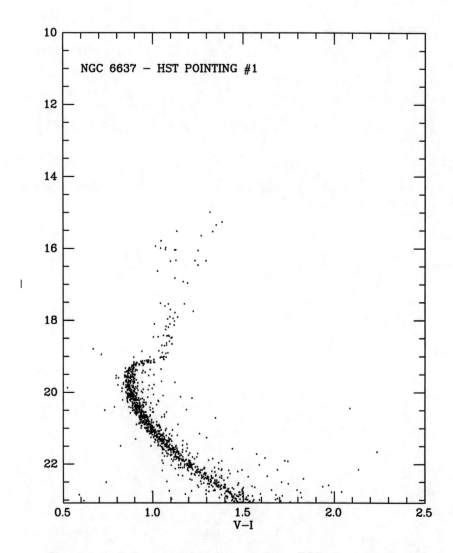

Figure 8. Color-magnitude diagram from *HST* observations of the outer, less sparse region of NGC 6637.

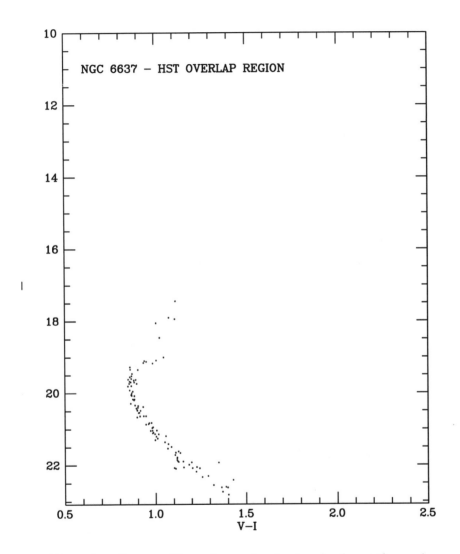

Figure 9. Same as Figure 8 except only stars in the overlap region between the two *HST* pointings are plotted.

Increasing Precision and Accuracy in Photometric Measurements

Michael V. Newberry

Axiom Research, Inc., Tucson, Arizona

1. Introduction

There is a myriad of image processing and data reduction techniques in various levels of common knowledge. Proper application of these techniques can lead to accurate, high precision photometry. However, there are some little known limitations built into these methods which can become important when attempting to achieve "milli-mag" level photometry on brighter objects or the very best photometry on faint objects.

Each pixel of a CCD camera records incoming photons as electronic charge. This charge produces a signal containing random noise from many sources, not the least of which, is nature's gift of Poissonian noise. Poissonian noise results from the random arrival rate of photons and produces noise proportional to the square root of the signal. There are instrumental noise sources as well, related to a wide range of effects, including seeing, CCD readout noise, and background brightness (Newberry 1991). The goal of good observational technique is to minimize all such instrumental contributions, producing observations that are as close as possible to being noise limited by nature. The CCD camera gain and the background subtraction technique are two such issues that are generally ignored, but which contribute noise to the final result, moving photometric results away from the ideal of being limited only by nature. Understanding these issues allows us to minimize their effects and attain photometry of higher precision. In this paper, I will discuss these issues and offer remedies.

2. The Effect of the CCD Camera Gain

Perhaps one of the most obscure places to search for unnecessary noise in a photometric measurement is in the gain setting of the CCD camera. Yes the particular value of the gain does indeed add some amount of noise to the image, and this noise propagates into the photometric measurements made from that image. Before discussing this issue in detail, I will give a short overview of gain and how the gain is usually selected for a given camera.

Extracting an image, or "reading" the CCD uses electronic logic to quantize the charge of each pixel into a digital number, usually called the "count" in the pixel. Pixel quantization is performed by an Analog to Digital converter, which applies a scale factor to the electronic charge in order to resolve it into a range of values that can be handled by a computer. The scale factor is called the "gain" of the camera system. In doing precision astronomical photometry, it is important to understand both how the gain value is selected and how the gain value affects photometric measurements made using images the camera.

For most CCD cameras used in serious astronomical work, the signal is digitized into a 16 bit number, giving a count range of 0 through 65535 counts. However, fast readout speed and budgetary issues can constrain the astronomer to using a 14 and 12 bit CCD camera (as a side point note that, in practice, the full count range of a given number of bits is never achieved because of electronic design issues). The CCD itself has a certain charge capacity that is governed by its specific architecture and not the readout electronics. For example, one CCD may be able to store only 50,000 electrons (e-) of charge, whereas another design may be able to store 350,000 electrons of charge. The capacity of the CCD is known as the "full well capacity", or "well depth". By adjusting the gain of the camera system, we can match the well depth of the CCD to the count range of the camera electronics. For example, we may wish to scale 350,000e- into 65,000 counts (= 16 bits). Setting the gain value close to $350,000/65,000 = 5.4$ would do this nicely. This means that 1 count is created in the pixel of the digital image for every 5.4 electrons, on average, that are stored in the CCD pixel. Using 14 bit digitization with the same CCD, the full well depth of the CCD would be utilized by matching the well depth to approximately 16,000 counts, or a gain value of $350,000/16,000 = 21.9$. Unfortunately, the common practice of setting the gain so as to digitize the full capacity of the CCD has implications for the noise performance of the camera system. In general, the higher the gain, the higher the quantization noise.

The effect of digitization bits and gain on the noise can be easily shown by a concrete example in which a digital image is multiplied by the gain factor to determine the number of electrons in the original signal. This will show the extra noise added by coarseness in the quantization process. Let us assume that we are using a CCD having a full well depth of 130,000 e-, and that an object is to be measured over an area of 4 pixels. I will consider the cases for systems with gain values of 2, 8, and 32. These values might be used in systems that digitize to 16, 14, and 12 bits, respectively, with the gain set to digitize the full well depth. The calculations will consider a brightness of 100e- incident on each of 4 pixels. The signal recorded by the CCD is considered to include Poissonian noise equal to the square root of the signal. Table 1, below, lists the original signal in the CCD pixels, including noise, and the digitized signal we measure in the image. The digitized signal is obtained simply by dividing the incident signal by the gain factor. Table 2 lists the measured signal after being converted back to units of electrons by multiplying by the gain factor. As shown in Table 2, quantization loses information by chopping the incident signal level into coarser measurement units. Comparing the "reconstructed" signal with the original, incident signal, Table 3 summarizes the photometric errors incurred by the quantization process. The second to last row of Table 3 lists the "Normalized error", which is simply the relative error divided by 2. The Normalized error reflects what we would actually measure, as the measurement would be calibrated so that the systematically low after-quantization errors would be split in half. That is, the zero point is implicitly calculated so that half the count residuals are above 0 and half are below 0. The bottom row in Table 3 lists the photometric error incurred by digitization, which is calculated from the relative error, r, using the formula $\sigma(m) = 1.0857\ r$.

From Table 3, the detrimental effect of high gain values is clearly visible. Clearly, one would probably not attempt high precision photometry using a 12 bit CCD camera, but gain values as high as 5 to 6 are commonly used with SITe (formerly Tektronix) CCD's to utilize their 350,000e- or higher well depth.

Table 1. Input signal level and digitized counts for 3 gain values.

Pixel #	Signal	g=32	g=8	g=2
1	100 e-	3 counts	12 counts	50 counts
2	86	2	10	43
3	103	3	12	51
4	115	3	14	57
Total:	404	11	48	201

Table 2. Signal measured in counts converted back to units of electrons.

Pixel #	Exact	g=32	g=8	g=2
1	100 e-	96 e-	96 e-	100 e-
2	86	64	80	86
3	103	96	96	102
4	115	96	112	114
Total Recorded Flux	404	352	384	402

Table 3. Summary of measurements and errors.

Digitization	Exact	G=32	g=8	g=2
Recorded flux	404 e-	352e-	384e-	402e-
Relative error	-----	-13%	-5%	-0.5%
Normalized error	-----	6.5%	2.5%	-0.25%
Photometric error	0.000 mag	0.071 mag	0.027 mag	0.003 mag

Based on the concrete example above, it is useful to look at the theoretical basis of the quantization error. Quantization noise can be characterized by computing the second moment of the probability distribution. It is easy to see that the probability distribution is "flat", meaning that electrons are distributed uniformly within a count. Working through the second moment of this probability distribution gives the quantization noise in terms of the gain as

$$\sigma = \sqrt{\frac{g^2 - 1}{12}}$$

This equation has two interesting limits:

$$\sigma = 0, \ for \ g = 1, \ and \tag{1}$$
$$\sigma \approx 0.29\,g, \ for \ large \ g.$$

That gain=1 generates zero quantization noise is heuristic, but it is important to recognize that *any* gain g > 1 adds a base level of noise to the CCD image. It is also clear that the noise contributed by the gain increases as the gain increases. The quantization noise does not scale with either time or signal level, and, in that sense, it behaves like the familiar readout noise. In fact, measuring the readout noise of a CCD system implicitly includes the digitization noise. This suggests that simply reducing the gain can reduce the effective readout noise of the camera system. Of course, reducing the gain also reduces the CCD well depth that can be digitized into the available number of bits. But for precision photometry the noise is usually of greater importance than being able to digitize maximal signal levels. In the normally photon starved realm of spectroscopy, lowering the gain can have an even more significant impact on the quality of the data.

Finally, to summarize the effect of quantization noise, some models of magnitude error as a function of signal level were generated using the parameters listed below.

Model Definitions:
σ: Noise, in e-
R: Readout Noise, in e-
g: Camera Gain, in e-/count
S: Signal, in e-
n: Number of pixels occupied by point source
Quantization noise: $\sigma_q = \sqrt{\frac{g^2-1}{12}}$
Poissonian noise: $\sigma_p = \sqrt{S}$
Readout noise: R = 0, 8 e-, and 15 e-
Total noise: $\sigma^2 = S + n\left[(\frac{g^2-1}{12}) + R^2\right]$
Sky: Assume negligible (very dark and measured over many pixels).

The models in Figures 1 through 3 illustrate how the two constant noise sources, readout noise and quantization noise, contribute to the total noise budget of a photometric measurement. The zero readout noise case in Figure 1 shows only the affect of the quantization noise. Figures 2 and 3 show progressively higher readout noise in conjunction with various gain settings. In Figure 1, the lowest curve denoted by triangles shows the case for gain g=2 and readout noise R=0. This is almost the Poissonian noise limited case we would like to achieve. Addition of readout noise and quantization noise strongly affects measurements below $\sim$1000e-. In particular, for source detection near the generally accepted 3-σ level ($\sigma \sim 0.3$ mag), the effect is remarkable. At extremely low signal levels, the difference between detection and non-detection is strongly controlled by both readout noise and quantization noise.

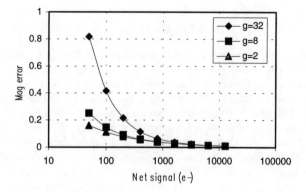

Figure 1. Models for R=0 and various gain values.

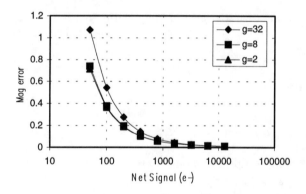

Figure 2. Models for R=8e- and various gain values.

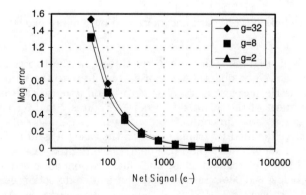

Figure 3. Models for R=15e- and various gain values.

3. Errors in Background Subtraction

Stellar photometry resolves to two issues: 1) integrating the count attributable only to the star, and 2) subtracting the background.

The generalized instrumental magnitude can be described in terms of a zero point constant k the net pixel counts summed over n pixels, which may be either partial or whole inside the aperture. If we let the signal attributable to the object be S_i, the signal attributable to the background be $S_{b,i}$, and the area of the i-th pixel be Ai, then the magnitude, m, of the star inside the aperture is given by

$$m = k - 2.5 \log \sum_{i=1}^{n} (A_i S_i - S_{b,i})$$

The difficulty in this equation is twofold: 1) determining the aperture area A_i, and 2) estimating the background $S_{b,i}$.

Integrating the count is straightforward after the area of the star has been defined. Usually, the star's footprint in the image is defined using a circular aperture, although an elliptical aperture is a more general case. Since photons define the signal, whereas noise attributable to sources beyond photons grows with the number of pixels under the object, measuring the greatest signal over the fewest pixels optimizes a photometric measurement. Since small tracking irregularities tend to smear the star image into an elliptical shape, this criterion is met by measuring stars through elliptical apertures. For small apertures, and especially with under-sampled data, the integration of light within the aperture involves computing the areas of numerous partial pixels along the perimeter. Along the perimeter, the intensity of each partial pixel is weighted by the fraction of pixel area enclosed by the aperture. Figure 4 shows a non-rotated elliptical aperture placed arbitrarily on the pixels of a CCD image.

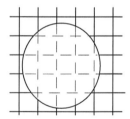

Figure 4. An elliptical photometry aperture placed on the pixels of a CCD.

An exact solution even for the area inside a circular aperture is quite complex to realize in a computer algorithm, but this difficulty is compounded when using elliptical apertures because an arbitrarily rotated ellipse is not a function. A generally oriented ellipse has two non-symmetrically placed y values for a given x coordinate, thus creating an algorithmic nightmare for the mathematician and programmer. Realizing small elliptical apertures is such a daunting problem that only the MIRA software package from Axiom Research, Inc. implements an exact solution for elliptical apertures. An elliptical or circular aperture area solution is easily checked by summing the partial pixel areas and comparing this sum with the value πab, where a and b are the ellipse semimajor and semiminor axes, respectively. The difference between an exact or approximate solution is usually negligible except when the aperture has a radius smaller than a few pixels.

Measuring the background is not a straightforward task. In this process, we seek to estimate the amount of signal in pixels under the star profile. Since we can never know the actual background, we increase the precision of the measurement using as many independent estimates of the background as possible, and use methods that lead to accurate estimation of the background brightness. Figure 5 shows an intensity cross section plot through a star profile, showing the actual data, the true point spread function (smooth curve), and 3 estimates for the background brightness. The background is represented by the term S_b in the equation above. Case 1 shows the background too low. Subtracting the Case 1 background from every pixel attributes too much light to the star, and we make an error in the magnitude estimate. Subtracting the high background of Case 3 leads to an underestimate of the signal attributable to the star, leading again to an error in the magnitude estimate. Case 2 shows the ideal background level we wish to subtract. When an ensemble of stars is measured, one would hope to alternate between the various cases so that, on average, we approximate case 2. This has the effect of converting a systematic error for each star to a quasi-random error for the ensemble of measurements. If we tend to measure case 1 or case 3 too often, then we realize a systematic error in the background measurement.

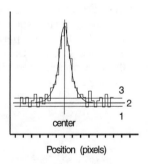

Figure 5. Background measurement showing an intensity slice through a star profile.

It is useful to estimate the effect of errors in the background subtraction. We can derive a simple equation that estimates the magnitude error based on the signal and a background error applied to every pixel under the object. The magnitude, m, is defined in terms of the net signal S, in the equation above. We take differentials of m and S to arrive at the general error relationship

$$dm = 1.0857 \frac{dS}{S}$$

Using the equation above, we can show how an error in the background propagates into an error in the magnitude:

$$dm = 1.0857 \frac{\sum_{i=0}^{n} dS_i A_i}{\sum_{i=1}^{n} (S_i - S_{b,i}) A_i}$$

The background error in each pixel is dS_i. Let us consider two examples in which the background is in error by the so-called "1-sigma" amount. That is, We estimate the sky by some statistical means, such as a least squares fit and consider the result to be above or below the true background level by the 1-sigma amount. In fact, probability theory says that we have only a 68% probability of our estimate being within 1-sigma of the true, but the background level is unknown, so this is not an unreasonable example.

Let the star have a total signal of 500 counts above the sky using a CCD camera with a gain of 2e-/count. Let the nominal background brightness be S_b = 400 counts. The Poissonian noise in the background brightness for 1 pixel is $\sqrt{400 \times 2} \approx 28$ e-, or 14 counts. Assume the background is estimated over 100 pixels. This reduces the error of the mean background to $dS_b = 14/\sqrt{100}$ = 1.4 counts. Now let us consider how this background error affects the star's magnitude measurement under astronomical seeing conditions that are good or mediocre. These results consider only the error contributed by Poissonian noise in the background fit, and do not include readout noise, or other noise sources in the background or object measurement. However, note that the process of fitting of the background implicitly accounts for all noise sources present in the background pixels.

Example 1: Good seeing.

Measure the star in an aperture of radius 3 pixels, hence an area of 30 pixels. The 1-sigma background error propagates into a magnitude error of

$$dm = 1.0857(1.4 \times 30/500) = 0.09 \, mag.$$

Example 2: Mediocre seeing, poor focus, or poor guiding.

Each of these factors has the same effect of enlarging the star to cover more pixels that in example 1. Measure the star in an aperture of radius of 5 pixels, or an area of 80 pixels. In this case, the 1-sigma background error becomes

$$dm = 1.0857(1.4 \times 80/500) = 0.22 \, mag.$$

One strategy might be to "beat down" the background noise by using more pixels in the background estimate. For example, increasing the background sample from 100 pixels to 10,000 pixels reduces the 1-sigma background error from 1.4 counts to 0.14 counts. However, this strategy has two pitfalls:

1. The larger the area used in fitting the background, the more likely the background estimate is systematically wrong. This results from the fact

that pixels further from the star carry more weight in the fit and may not estimate the local background nearly as well as pixels closer to the object of interest.

2. If the image being measured is stored in an integer pixel format, such as 16 bit integers, quantization noise in the pixel values contributes to the total noise budget (see section 2). Specifically, 1 count adds a quantization noise of 0.29 counts to the data.

Item 2 shows that one reaches the quantization limit well before reaching the 1-sigma level of 0.14 counts even if 10,000 pixels are used in estimating the background. Whenever an inherently real quantity it converted to an integer, one unit of quantization noise is added to the image in a root sum square fashion. Therefore each stage of data reduction that produces inherently real valued pixels, such as flat field correction, or overscan bias correction, adds quantization noise when the result is saved as an integer type image. The solution to this dilemma is to process and store images in real pixel format, for example, using 32 bit real pixels.

4. Conclusions

In this paper I have presented some obscure noise issues that can be important for high precision photometry. In particular, one must be aware of quantization and noise propagation issues. Some specific recommendations are as follows: 1) Use a CCD camera with gain as close to 1 as possible, although a gain value as high as 2 gives good results. Also, use a 16 bit camera when possible. 2) Convert images from integer to real pixel format as early as possible in the data reduction process. 3) When doing aperture photometry, elliptical apertures are preferred over circular apertures. Especially when working with data that are undersampled or were acquired in excellent seeing, use an algorithm that provides an exact solution for the aperture area.

References

Newberry, M. V. 1992, in Astronomical Data Analysis Software and Systems I, A.S.P. Conference Series Vol. 25, eds. Diana M. Worrall, Chris Biemesderfer, and Jeannette Barnes (San Francisco: ASP), 307

Newberry, M. V. 1991, Publication Astr. Soc. Pac., 103, 122

The GNAT Automatic Imaging Telescope for CCD Photometry

Eric R. Craine

Western Research Company, Inc., Tucson, AZ 85719, USA/GNAT, Inc., 2127 E. Speedway, Tucson, AZ 85719, USA/Department of Physics, Colorado State University, Ft. Collins, CO 80523, USA

David L. Crawford

National Optical Astronomy Observatory, Tucson, AZ 85721, USA, GNAT, Inc., 2127 E. Speedway, Tucson, AZ 85719, USA

Patrick R. Craine

School of Engineering, University of California, Berkeley, CA 94709, USA, GNAT, Inc., 2127 E. Speedway, Tucson, AZ 85719, USA

Abstract. The Automatic Imaging Telescope (AIT) is now a reality which provides a powerful tool for very efficient, and potentially high precision, CCD photometry. Such telescopes can be organized into distributed networks of telescopes, such as the concept advanced by the Global Network of Astronomical Telescopes (GNAT), to attain added power derived from greater versatility. AITs have peculiar practical limitations which impact on photometric precision to be generally attained with such systems, and many of these limitations are the subjects of ongoing study. We report here on the status and continuing development of AIT tools underway for GNAT.

1. The Global Network of Astronomical Telescopes

There have been many improvements in telescope technology over the last few decades. We now have light weight mirror blanks, active optical surface and wavefront control, fast focal ratios for the primary, smaller and lighter weight mountings, full usage of computers in telescope control and data handling, smaller and thermally more friendly housings, and many other items.

It is important to note that many of the items listed above also come into play in the design and operation of small telescopes. As with the larger ones, it also means that costs are less for a given aperture than in the older generation telescopes. While the costs of telescopes scale with some relatively high exponent of the aperture (about 2.6, for example, "all things being equal"), the costs for telescopes of all apertures are much lower (in inflation adjusted currency) than previously. We are now seeing computer controlled, automatic and remotely operated small telescopes in routine operation at a number of observing sites.

The technology is still changing rapidly, and it is sure to bode well for both large and small telescopes.

In addition, the state of art in astronomical instrumentation has improved greatly in recent years. An imaging CCD photometer is a powerful tool for a small telescope, making a one meter aperture telescope more powerful for many research problems than the 5-m telescope was only a few decades ago.

Another change in technology that has been of great benefit for astronomy has been the improvements in electronic communications, e-mail and the Internet and the World Wide Web, for example. We now can communicate rapidly with our colleagues most anywhere, sending data back and forth and accessing available data bases in a very efficient manner. We can also control our new generation telescopes remotely in addition to getting the data from them.

It is easy to envisage, therefore, a global network of automatic small telescopes, operated remotely from a "Homebase." Such telescopes would be very powerful tools for many important problems in astronomy, such as monitoring all sorts of variable objects, from quasars and lensing galaxies to nearby stars, and even the planets and asteroids. They would be most valuable for many sorts of surveys. They would be a critical resource for improvement in photometric systems and standard stars. Finally, they would be able to supply a good deal of observing time for many excellent scientific programs for which there are simply not enough resources otherwise available.

GNAT, Inc. is a non-profit research and education corporation with a membership of both institutions and individuals (Crawford 1995; Crawford and Craine 1996). The GNAT charter is carried out by the use of a Working Group structure, with each Working Group having responsibility for a specific aspect of the GNAT tasks. Examples of active Working Groups at present are: Telescope Development, Instrumentation, Telescope Control Systems, Standard Stars and Standard Photometric Systems, High Energy Source Monitoring, Science Education Outreach, Funding, etc.

2. The Prototype Telescope

2.1. The Telescope Design

An early effort of GNAT was to explore the commercial availability of reliable automated telescopes. This effort was undertaken because the GNAT charter was to design a network for the acquisition of astronomical data, not to become a manufacturer of telescopes. Further, activity in the development of automatic telescopes led to the widespread belief that such systems were readily available. In April 1995, GNAT hosted a meeting in Tucson, Arizona to discuss requirements for a GNAT telescope which would meet research and education needs of the GNAT membership; a summary of the resultant requirements can be found on the GNAT web site (Web 1998b). The search for a vendor who could satisfy these requirements quickly established that the commercial availability of automated telescopes had been greatly overstated (Sinnott 1996); popularly advertised systems either did not exist as deliverable products, or they were extremely expensive and not clearly demonstrated as automated telescopes.

An agreement was eventually struck with SciTech Astronomical Research (Sacramento, CA) to explore a possible telescope design proposed by that company. The plan was to install an existing half scale (0.5-m aperture) model of the SciTech telescope at a site in Tucson, Arizona to enable testing of the design and development of the software required of the automated GNAT telescope. If the design were found satisfactory, it would become the basis for the 1.0-m aperture telescopes to be used in the primary GNAT network.

The prototype 0.5-m telescope which is currently in use as the GNAT testbed instrument is an f/5.6 e.f.l. Ritchey-Chretien system. Focus control is provided by a single digital servo motor serving three independently adjustable focus position ball screws to adjust the secondary mirror position. The optics are supported in an open frame optical tube assembly. The telescope mounting is an equatorial yoke fabricated from sections of a steel sphere which are welded in a very rigid monocoque structure. A stepper motor roller drive system operates on large radius fabricated drive bands, each calibrated with an attached laser engraved position encoder fiducial strip.

The instrument bay for this telescope is a hemispherical tub located immediately below, and partially enclosing, the primary mirror. Access to the interior of the tub is via two hatches just behind the declination axis. During the test and evaluation phase of this program the telescope was modified with the addition of a computer controlled filter changer and a CCD camera based autoguider. The filter changer is a stepper motor controlled wheel with space for six 2x2-in filters (this number constrained by the very limited space available in the instrument bay. The filters in current use are parfocal U, B, V, R, I and Z band filters; the U-I filters based on designs of Mike Bessell (Bessell 1996). The autoguider is a modular unit which uses a small pickoff mirror mounted to rotate through an annulus just outside the main beam of the telescope which falls on the science camera. Opposite the pickoff mirror, and mounted on the same chassis, is an Electrim E1000 CCD camera. The autoguider will be controlled by the GNAT Telescope Control System (GNATCS) to automatically locate and track a guide star in the annulus near each program object.

The general layout of the GNAT 0.5-m prototype is shown in the drawings of Figure 1. This telescope is planned to be scaled up to a 1.0-m telescope as the primary system for the full scale GNAT system. Although some changes will be made based on our experiences with the prototype half-size scale model, the general appearance of the 1.0-m telescope is expected to be very like that of the 0.5-m.

At this writing there are several observing programs underway for the use of various GNAT members. All of these programs have been chosen to provide simultaneous engineering data useful in continuing to evaluate and improve the system. There remain aspects of the telescope which must be implemented, for example the autoguider system is not yet operational, an automatic focuser is not in place, and the control software is still in the active development phase. In addition, as we operate the telescope for the early stage science programs, we are taking note of various deficiencies or inconveniences which must ultimately be addressed. Most, if not all, of these engineering issues can and often do have a profound impact on the resultant photometric precision in the science data.

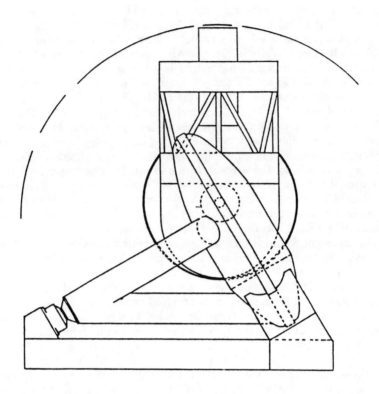

Figure 1. General layout drawings of the GNAT 0.5-m prototype AIT.

2.2. GNAT Telescope Control System (GNATCS)

Initial telescope operations were undertaken using the SciTech telescope control system. This prototype software was not yet complete and was not structured for fully automatic operation. A decision was made to develop an appropriate control system within GNAT to result in a complete, automatic controller for the telescope, instrumentation and scheduling. The resulting system is the GNATCS software, which is presently nearing completion. First successful full automatic operation of the prototype telescope under GNATCS was accomplished in January 1998, although numerous improvements to the software have been required to accomplish moderately routine, full night automatic operations. This goal was reached in the Fall of 1998.

GNATCS is a Linux based package written in C; it is presently run on a Gateway 2000 133MHz Pentium machine serving as the telescope host controller. The user interface is an X-windows display comprising several windows which may be selected for viewing during manual telescope operation. The basic window structures, and associated functions, are as listed:

```
MAIN----|---- paddle
             |-------      set
             |-------      slew
        |---- coordinates
             |-------      manual input of target coordinates
             |-------      precession
        |---- CCD camera
             |-------      automatic operation
                          |-------   links camera to ATIS control
             |-------      manual operation
                          |-------   select image type
                          |-------   select filter
                          |-------   input header information
                          |-------   select image type (object, dark etc)
                          |-------   set integration time
                          |-------   automatic or manual file naming
                          |-------   start integration
        |---- CCD auto-guide
             |-------
             |-------
        |---- pointing calibration
             |-------      (this set of routines collects mount
                           model data)
        |---- virtual auto-guide
             |-------      CAL ON implements automatic pointing corrections
        |---- initialize
             |-------      sets telescope encoders for home position
        |---- misc  (future use)
        |---- ATIS operation
             |-------      open roof command
             |-------      read ATIS file
             |-------      START automatic observing program
             |-------      manual or automatic QUIT
        |---- focus
        |---- move
             |-------      manual move to selected coordinates
        |---- home
             |-------      moves telescope to home position
```

In its current mode of operation the telescope start-up tasks are performed manually. After initialization the automatic observing function is turned on and allowed to run throughout the night, at which point manual data downloading and telescope shutdown are performed. These operations will eventually also be automated.

The start-up tasks are as follows: 1) the roof is opened and power is applied to the system and the camera is turned on to start its cool-down process, 2) the telescope control system clocks are set and verified, 3) a set of twilight flats is obtained through each filter which will be used for the night, 4) a set of dark and bias frames is obtained, 5)) the telescope position is initialized, 6) the science camera computer is initialized, 7) the system initialization is confirmed by pointing to a known star and verifying its position in an image obtained by the science camera (also confirming that the camera is focused and working properly), and 8) the automatic control software is initialized and started.

At this point all operation of the telescope is given over to automatic control. Telescope pointing, filter changing, and image acquisition will continue, driven by the constantly updating logic of the controller, and ensuring observation of each object based on a user-programmed set of priorities, until such time as there are either no more observations to be made, or until astronomical twilight occurs.

The shut-down tasks are as follows: 1) obtain any additional manual calibration data desired, 2) download the image data, 3) shut down the science camera system, 4) confirm that the telescope has automatically gone to its home position, 5) download the observing logs, 6) shut down the GNATCS system, and 7) close the roof and secure the facility.

2.3. Automatic Telescope Instruction Set (ATIS)

Automatic operation of the GNAT prototype telescope is accomplished by using the Automatic Telescope Instruction Set (ATIS). ATIS was developed originally for purposes of providing observing program instructions to automated photoelectric telescopes (APTs). GNAT requirements add several additional dimensions to automated astronomical observation, primarily because the mode of observation is imaging, and because the goal is to cause a large number of telescopes, at widely separated sites, to work efficiently in concert. An overview of ATIS93, the current version of ATIS, can be found in several sources, but background descriptions and software templates are most conveniently downloaded from the web (Web 1998a).

ATIS is a large number of command formats from which may be selected specific commands which can enable a specific observing program. The commands are combined in a text file such that the final observing program can be read, and then acted upon, by the control system, in our case GNATCS. Until recently, ATIS files were prepared in a word processor or editor. A convenient, windows driven ATIS file creator, called XCreate, has now been developed by Don Epand. XCreate allows construction of the older ATIS command streams for APT photometers, but also for the newer imaging systems, the AITs.

An example of a part of a minimal GNAT ATIS file for operation of an AIT is shown below:

```
103
1 10 1 2450735 2451544 4.975 5.975 1 10 100 3 RXBoo Scharlach/Craine
104
1 RXBoo 1 0 Mira
105
14 24 11.6 25 42 13 2000
510
1 3 1 0 20.0 1 1 0.0 0.0 1 0 0 0 0 1 0 RXBoo
510
1 5 1 0 20.0 1 1 0.0 0.0 1 0 0 0 0 1 0 RXBoo
115
```

This ATIS command string represents one group of observations and consists of five distinct ATIS commands. Each command is identified with a single three digit number, in this example 103, 104, 105, 510 and 115. The line after the command identifier contains the actual instructions for that specific command. These commands serve the following functions: 103 - Group Header, with group identification and selection criteria, contains an identifying serial number (1), starting and ending Julian dates, times of observational window, number of observations, priority, and group name and observers' names; 104 - Star information, passes information on the nature and identification of the object to be observed; 105 - Move command, instructs the telescope to move to a position specified by celestial coordinates and epoch for the object to be observed; 510 - Take Image, passes instructions to the CCD camera host computer, and contains information on the desired filter (in this case V=3 and R=4), the integration time (20 seconds) and various other desired camera parameters; and, finally, 115 - End of Group.

If this group represented the entire ATIS file, then invoking the ATIS Operation command from the main GNATCS window would cause this file to be read, at which point, if the Mira variable RX Boo were in an accessible part of the sky (as defined in the ATIS file), the telescope would immediately slew to the object coordinates, insert the V filter in the optical path, initialize the CCD camera, clear the charge from the CCD chip, execute a 20 second integration, read the image to the camera host hard drive, insert the R filter in the optical path, repeat the imaging process, and then automatically quit.

2.4. CCD Camera

The CCD camera used in the GNAT prototype system development is from Southwest Cryostatics (Tucson, AZ) and has been designed and built by GNAT member Roy A. Tucker. This camera is of particular interest to us since the cryostat design, as well as the support electronics, enable easy replacement of the detector both as upgrades to the camera, and as a testbed for different types of detectors. The camera presently in use with the GNAT system uses a SITe TK512 CCD array, although the same camera has also been used with a SITe 1024 array.

The detector is mounted in the cryostat on a pedestal comprising four thermoelectric coolers in series. A vacuum valve located on the back plate of the cryostat enables it to be periodically evacuated using a simple two-stage mechanical pump. A dry ice cold trap has recently been replaced by a molecular sieve, and a series of tests of camera cooling as a function of time and cryostat evacuation has been undertaken (Craine 1996). With a freshly evacuated cryostat the camera will operate at a temperature of -60F; the temperature is held constant to $< 0.1°$ for periods of 20 hours or longer, thus making feasible CCD photometry with some reliability. Thermoelectric cooling is necessary at this stage as the telescope will soon be operated unattended for several days at a time. A discussion of recent experiments with the cooling of this camera has been prepared by Patrick Craine and appears elsewhere in this volume.

The CCD camera is controlled by a separate computer from the host computer for the GNATCS, although the two communicate via a serial port link. The CCD controller, DOSCAM, is a DOS based system resident in a 486 PC, which receives the outputs of ATIS commands from the GNATCS computer. The CCD controller computer has a 3 Gb hard drive and a Zip drive for downloading images. One to a few nights observations can be stored on the hard drive then written to one or more Zip disks for transport to Principal Astronomers, students running projects on the telescope, or to the GNAT offices for subsequent processing. We also maintain a CD-RW capability, although we have become reluctant to make that a primary mode of archiving images as a result of recent reports of temporal failures in CD media resulting in the loss of observing data.

2.5. Development Observatory Facility

The GNAT prototype telescope is undergoing test and further development in a small roll-roof observatory located in Tucson, Arizona. The observatory is 12x30-ft with an attached workshop and storage area; the telescope is housed in a central 12x15-ft area with a double clamshell sliding roof, suitable for a single telescope. The GNAT observatories which will house the 1.0-m telescopes will be of similar design, but with a sufficiently large pad to accommodate four telescopes at a single site.

Basic weather data is derived from a commercial weather station (Davis Instruments, Hayward, CA) which can be directly logged into a 486-PC. Weather sensors for rain and cloud have not yet been installed, nor has an automatic roof closing mechanism. A two stage model 1405 Duoseal vacuum pump (Welch Vacuum Technology, Inc., Skokie, IL) is mounted under the telescope pier for easy access to the CCD cryostat. Flat fielding is enabled by installation of a spectrally flat screen (SORIC, Thornton, Ontario, Canada) at the south end of the building. This installation is not trivial due to the small size of the observatory and a plastic protective housing was fabricated for protection of the screen from periodic traffic in the building, as well as isolating humans from the potential health hazards of the silicone fiber screen. A retraction mechanism allows the screen to be swung into position for use and then safely stowed after "dome" flat fields are obtained. Lights for flat fields are mounted on the upper ring of the optical tube assembly.

2.6. Current Observing Programs

The GNAT 0.5-m telescope is intended initially as a test system to help refine the GNAT 1.0-m telescope designs (see paper by Jim Wray in this volume). However, a part of that process involves experimenting with the 0.5-m through a variety of observing programs, most of which are now in early stages of data collection. The observing programs are of four types: 1) engineering, 2) photometric precision, 3) research and 4) student projects.

The engineering programs are of highest priority at present. These observing sessions consist primarily of automated ATIS observing runs designed to collect data necessary to improve the telescope mount model, to serve as input data to analysis programs for detailed characterization of the telescope hardware and to address issues related to manufacturing processes for the 1.0-m telescopes. These programs are also directed at exhaustive exercising of the GNATCS system to search for errors and to help define the best mode of operation of the overall system. They will also involve tests to evaluate the use of GNATCS in a distributed network of GNAT telescopes.

At this writing the GNATCS software is best defined as existing in a Beta-test mode, that is it is functional, but not exhaustively tested. We have been successful at operating the telescope in an automated mode, but the system is

Figure 2. The prototype GNAT 0.5-m telescope seen shortly after temporary installation at a GNAT test observatory in Tucson, Arizona. This telescope has been used as the test bed for GNAT software development, acquisition of data for student programs and initial exploration of the photometric precision attainable with the proposed GNAT systems. The telescope is now operated automatically using a control system developed by Don Epand and will serve to help refine design issues for the GNAT 1.0-m imaging photometry telescopes.

still rather "buggy" which is most frequently manifest in the form of improper shutdowns. The result is that we are able, on most nights, to collect useful images, but the system does not always behave in the preferred manner, and occasionally fails altogether. Nonetheless, these are valuable occurrences which enable us to refine the software until we achieve a robust system which will work consistently and reliably.

The next highest priority is evaluation of photometric precision attainable with the system. This is the subject matter of one of the most active of the GNAT Working Groups and is critical information to develop in preparation to determining the types of science programs which can be most successfully undertaken with the various GNAT facilities. This activity will also become an important part of the 1.0-m telescope system design as we strive to provide the best possible automated network for imaging photometry.

To this end we have set up several photometric reduction facilities, all Pentium PC based. At present we are in different stages of gaining experience with three software tools, all further discussed elsewhere in this volume. These packages are IRAF, SPS and MIRA. We are in the process of reducing selected GNAT image sets using all three of these software tools and plan to invest a major effort in understanding the sources of errors which are inherent in the GNAT systems. Some of these data reduction experiences are recounted in other papers in this volume.

Several GNAT members have research projects which will be supported on the 0.5-m telescope during a modest research effort over the next two years. These include projects to refine photometric standards, search for close companions to Mira variable stars, search for previously undiscovered variable stars in open clusters of known distance, monitor the brightnesses of possible binary asteroids, and monitor late-type stars which are potential hosts of extra-solar system planets.

The most active of these examples at this stage is a project by Werner Scharlach to investigate light curves of selected Mira stars in order to explore multiplicity of the star systems. The observations are made with both V and I band filters, typically with 30 to 180 second integrations. Since the autoguiding system is not yet installed, this program works well for us right now as the telescope tracks well for up to 240 seconds, when some trailing can start to become apparent. Twilight flats in each filter, darks and bias frames are typically made at the beginning of the night. At present these are done manually, although we will soon incorporate these observations in the ATIS files as well. When the sun angle reaches 102° the ATIS file can be started. The telescope then automatically observes each program group, ideally until it reaches a termination command or runs out of program fields to observe. A group can typically execute in about 320 seconds, and new field acquisition usually occurs in about 15 seconds (since the ATIS file is optimized to observe groups close together and within about 20 minutes of the prime meridian), thus we typically make multi-color observations of about 12 program stars per hour, or usually 100-120 stars per night in this program.

Once this observing program smoothes out, and the telescope is fully operational, it is very clear that the real bottleneck will be the data reduction. Already we find it difficult to keep up with even partial nights of observing. The great efficiency and relentless objectivity of an automated observing system for purposes of photometry can quickly outrun a human observer in terms of data stream productivity, not only in terms of quantity but quality as well.

For the Mira star program, we return the raw images to our Tucson offices on Zip disks where the data are reduced, at present using MIRA routines. The instrumental magnitudes are extracted using both aperture photometry and growth curve photometry techniques. All of this work is using differential photometry and the program star light curves are constructed using ensemble means of the comparison stars.

Finally, GNAT is beginning to expand an effort to support student projects, at both high school and university undergraduate level, using data from the 0.5-m telescope. Some early efforts have already been completed (c.f. Culver and Craine 1996 as well as Jacob Taylor's paper in this volume); other projects are underway. For example, the Department of Physics at Colorado State University currently has several students, both undergraduate and graduate, enrolled in independent studies in astronomy projects using the GNAT 0.5-m telescope; at present these primarily involve studies of both open and globular star clusters.

The Colorado State University experience is of particular interest as it serves as a prototype for the typical GNAT node presently envisioned. The facility required on the university campus is modest and requires relatively little institutional commitment. The CSU site consists of some space provided by the physics program, and a computer (or computers) equipped for image processing. These are typically Pentium PC multimedia machines with sufficient storage capacity to handle a few hundred images at a time. Operating systems are usually Windows 95 and/or Linux. The Linux machine runs IRAF and a version of SPS; the Windows machine runs MIRA and another version of SPS.

Observing program planning can be done on campus, with the objects to be observed, and the mode and circumstances of observation, determined by the student. These input data can then be used to create an ATIS file, using the program XCreate, or they can be forwarded directly to GNAT headquarters in Tucson. At that point the observing program is reviewed and checked and then integrated into a master GNAT ATIS file, which is then loaded on the telescope controller. At the end of a night of observing images are now transferred to Zip disks, copied, archived, and distributed to the observer at his home institution.

Each observer receives only raw, uncalibrated images, so that all of the processing is left to the student. All of the processing can be done comfortably and conveniently at the home institution. This type of node allows the student to participate in all relevant phases of modern automated observing, to obtain high quality data on request, and to have the capacity to use front line reduction tools to process and analyze the data. The advantage to the university is an easy to support and maintain facility and an interesting and useful learning experience for the student.

3. Current Issues in Precision CCD Photometry

It is apparent that much of the effort devoted to GNAT so far has involved wrestling with hardware and software issues to enable the demonstration of a workable AIT which serves as a viable prototype system and a foundation for making future plans. We are now at a stage where, although the development work continues, the exploration of the precision with which photometric measurements can be obtained with the system is assuming great importance. The significance of this effort is that it will allow determination of the specific scientific programs, among the many potentials, which can be most effectively undertaken with GNAT.

It should be noted immediately that there are some inherent limitations peculiar to the operational and structural requirements of GNAT. Perhaps foremost is that in a practical sense GNAT will for some time be limited to telescopes of aperture on order 1.0-m, thus setting an upper limit on the number of photons which can be collected as well as an absolute limit on the signal-to-noise attainable for a given integration time. A second practical limitation is associated with the requirement that GNAT telescopes operate unattended , sometimes for extended periods of time. The result is that GNAT is, for the present, limited to CCD cameras which are thermoelectrically cooled. The camera in use with the 0.5-m prototype telescope has the benefit of four stages of cooling, which achieves a stable plateau of background noise, but still at a level much higher than attainable with a nitrogen cooled system. Discussions of this type of cooling can be found in this volume in papers by Roy Tucker and Patrick Craine.

4. Acknowledgments

The authors would like to acknowledge the substantial contributions of the following GNAT members: Don Epand, who has been instrumental in the development of the GNATCS telescope controller and telescope automation, Roy A. Tucker, who designed, built and has helped support the GNAT prototype CCD cameras, Jim Wray, who designed and built the GNAT 0.5-m prototype telescope, and who has designed the GNAT 1.0-m telescope, and Werner Scharlach and Michael Snowden, who have provided valuable feedback on test programs of CCD photometry. We also gratefully acknowledge the technical, moral and financial support of all of the GNAT members, institutional and individual, who are helping to make GNAT a reality. PRC has worked as a student intern assisting in various aspects of the telescope system installation, operation and testing.

References

Bessell, M. 1996, private communication with David L. Crawford

Craine, P.R. 1996, GNAT Tech.Reps., TR96-1101, GNAT:Tucson, AZ, 1.

Crawford, D.L. 1995, Robotic Observatories,ed. Michael F. Bode, John Wiley & Sons, New York, 77.

Crawford, D.L. and Craine, E.R. 1996, J. Roy. Astron. Soc. Can., 90, 224.

Culver, R.B. and Craine, P.R. 1996, I.A.P.P.P. Comm., 68, 49.

Sinnott, Roger W. 1996, S&T, June,38.

Web 1998a, http://24.1.225.36/atis.html.

Web 1998b, http://www.gnat.org, (see Past Meetings).

Observing Blazars with the WEB Telescope

J. R. Mattox

Boston University, on behalf of the WEBT collaboration

Abstract. The optical monitoring of blazars in conjunction with multiwavelength campaigns requires extensive observations at multiple longitudes. Automatic telescopes (such as those proposed by GNAT) would be very useful for this. An initial effort to coordinate blazar observations with telescopes at diverse locations is called the **WEB Telescope** (http://gamma.bu.edu/webt/).

1. Introduction

The blazar class of AGN includes objects classified as BL Lacertae type objects, high polarization quasars (HPQ), and optical violently variable (OVV) quasars. These sources feature strong, compact, flat-spectrum radio emission. They also display continuum domination of the optical emission, significant optical polarization, and often significant changes in optical flux on short time scales — for example, Figure 1 shows dramatic intra-night variability of THE BL Lacertae source (the eponym for this type of source).

The measured redshifts ($z \gtrsim 1$) of blazars correspond to large distances. This means that the observed flux densities imply blazar optical luminosities in excess of $\sim 10^{47}$ erg/s for some blazars if the emission is isotropic. It is difficult to understand a luminosity this large in a region as compact as that implied by the $\sim$1 hour variability time scale of many blazars. For this and other reasons (such as apparent superluminal motion of parsec scale radio structure), it is believed that the observed properties of blazars result from emission by material in a relativistic jet (with bulk Lorentz factors $\sim$10) which is directed within $\sim$10° of our line of sight (Antonucci 1993). The origin of these relativistic jets is one of the great mysteries of our times.

1.1. The Potential of Multiwavelength Astronomy

With substantial accretion onto a $\sim 10^8$ M$_\odot$ blackhole at the base of the jet, there is certainly sufficient power to create the jet. However, the physics involved in the conversion of gravitational potential to kinetic luminosity is not yet understood, and is likely to be very interesting. We believe that the continuum emission of these jets may eventually lead to an understanding of the origin of the jet. However, we still have a great deal to learn about the nature of the jet (e.g. are jets comprised of leptonic, or hadronic plasma) and the jet emission processes. The EGRET γ-ray telescope aboard the *Compton Observatory* has detected $\sim$1 GeV emission from $\sim$50 blazars (Mattox et al. 1997a). The apparent γ-ray

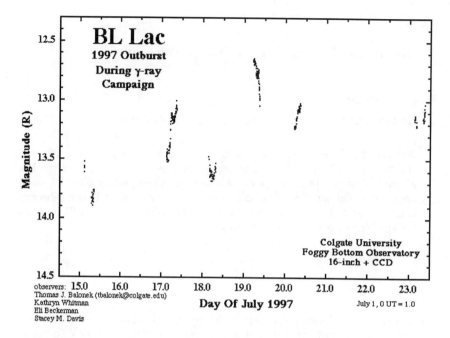

Figure 1. Optical variability of BL LAC observed by the group of Thomas Balonek during a flare in 1997.

luminosity seen for the EGRET blazars is as much as one hundred times larger than that at all other wavelengths for some flaring EGRET blazars. Variability of the γ-ray flux from some blazars on a time-scale as short as 4 hours has been observed (Mattox et al. 1997b). This implies that the region of γ-ray emission must be very compact. Because the opacity to $\gamma - \gamma$ pair production must be low for the γ-rays to escape, relativistic beaming with a Lorentz factor of $\sim$10 for the bulk of material in the jet is required (assuming that x-rays originate in the same volume as the γ-rays).

It is widely believed that this GeV emission is due to inverse Compton scattering by shock accelerated leptons within the relativistic jet. However, there is wide disagreement over the origin of the $\sim$1 eV photons which are scattered. It has been suggested that they might originate within the jet as synchrotron emission (Bloom & Marscher 1993). This is designated as the synchrotron self-Compton (SSC) process. Another possibility is that the low energy photons come from outside of the jet. This is designated as the external Compton scattering (ECS) process. Dermer, Schlickeiser, & Mastichiadis (1992) suggested that they come directly from an accretion disk around a blackhole at the base of the jet. It was subsequently proposed that the dominant source of the low energy photons for scattering could be due to re-processing of disk emission by broad emission line clouds (Sikora, Begelman, & Rees 1994).

The correlation of optical, x-ray, and γ-ray emission is expected to provide a definitive test of these models. If the Compton-scattered flux follows the synchrotron flux closely, the SSC is indicated. If the variation in the synchrotron

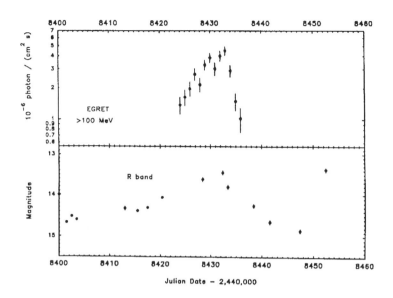

Figure 2. Simultaneous EGRET and ground-based optical measurements of 3C 279 in June of 1991 (Hartman et al. 1996).

flux is due to a change in the electron density, the SSC emission, which depends on the second power of electron density, will be observed to vary quadratically in comparison to the synchrotron. If the high-energy emission is ECS, it will lag the optical disk flux by $\sim 10^3$ seconds. If the disk photons are reprocessed in the broad line region before inverse Compton scattering (Sikora Begelman and Rees 1994), the ECS flux will lag the optical disk flux by at least 1 day if the ECS flare is due to increased optical emission from the disk. If the ECS flare is due to an enhancement in the relativistic particle content of the jet, no increase in the disk emission is required, but enhanced synchrotron emission should be apparent. Thus, multiwavelength observations offer the opportunity to understand the continuum emission of blazar jets. As we gain understanding of this emission, it potentially can then be used to study the origin of the jet.

In addition to providing for discrimination between models, simultaneous multiwavelength observations also have the potential to determine of properties of the jet; e.g., Takahashi et al. (1996) infer the strength of the magnetic field in the jet of blazar Mrk 421 by examining the rate of Synchrotron energy losses with the ASCA satellite.

Unfortunately, existing multiwavelength data are not adequate yet to permit definitive conclusions to be drawn about the nature of blazar jets. In most previous multiwavelength campaigns, the optical coverage by ground based observers has been much too sparse. Figure 2 shows such a result. There is apparently some correlation of variability, but a detailed study of relative flux change and possible temporal lag is not possible.

One of the best extant multiwavelength datasets for a blazar flare was obtained in the summer of 1997. This serendipitous multiwavelength study of BL LAC was initiated by the discovery of a major optical flare (see Figure 1). Simultaneous exposures by CGRO, RXTE, and ASCA were arranged. A large GeV flux was detected (Bloom et al. 1997). The ASCA satellite dedicated 24 hours to this campaign, a major amount of time for this observatory. We arranged for as much ground-based optical coverage during this 24 hour interval as feasible. The resulting optical data is shown in conjunction with the ASCA data in Figure 3. There were a few other optical observations of questionable calibration during this interval which are not shown.

Although the pending 24 hour ASCA observation was made known well in advance of the observation to all known optical observers of blazars, optical observations were obtained for only ∼5 hours out of 24. A double maximum is apparent in the X-ray data. There is not sufficient optical coverage to determine if this pattern is also in the optical emission. There is possibly correlation between the X-ray and optical variability, but the lack of optical coverage prevents a definitive conclusion from being made about the extent of correlation, and about lags. This lack of optical coverage is very unfortunate. Because of the complex variation and spectral evolution of the x-ray flux, a very interesting result may have been obtained with continuous optical coverage. Thus, the potential to obtain definitive insight into the nature and creation of this blazar jet was not realized last summer.

2. Whole Earth Blazar (WEB) Telescope

We suggest that a coherent use of existing ground-based optical facilities can dramatically improve the quality of the optical data obtained during multiwavelength blazar campaigns. Although the optical data in Figure 3 are inadequate for a detailed study of the cross-correlation of variability between the optical and x-ray energy bands, the data during the ∼5 hours when observations were actually made are of high quality. This is encouraging. Without CCD cameras and computers, these data could not have been economically obtained. Moreover, it is apparent to us that a coherent effort by participants at ∼10 observatories could have produced continuous, high-quality, high-temporal-density, optical monitoring during this 24 hour ASCA observation. And, typical of coherent processes, the result could be far more valuable than the sum of individual efforts incoherently executed. The **WEB Telescope** was conceived following the failure to obtain more than 5 hours of optical exposure during this BL LAC campaign.

The WEB Telescope (WEBT) is a network of optical observers who in concert have the capability to obtain continuous, high-temporal-density, optical monitoring of blazars. Because of the longitude diversity of participating observers, it is feasible to obtain continuous optical observations, with observing activity moving from east to west around the world as the Earth rotates. The Internet home page is **http://gamma.bu.edu/webt/**. The locations of observatories currently participating are shown in Figure 4.

An important area where the **WEBT** could also be very useful is in further study of the reality of the correlation of radio and optical Intraday Variability (IDV) of blazars as reported by Quirrenbach et al. (1991) for BL Lac type

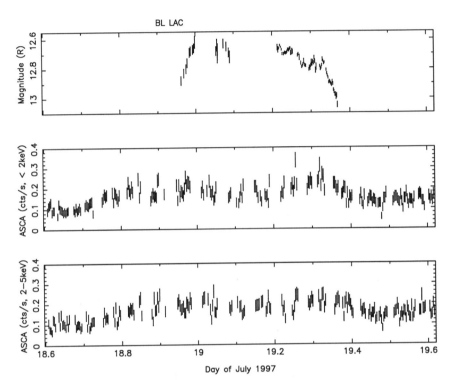

Figure 3. Ground-based optical and X-ray lightcurves for the 24 hours of ASCA observation of BL LAC during the 1997 flare. The ASCA lightcurve is given for two different energy bands because substantial spectral evolution is apparent (Tanihata et al. 98). The first segment of optical data was obtained by Gino Tosti using a 0.4 meter automatic telescope at Perugia Observatory. His observations were interrupted at midnight for ∼100 minutes by cloud cover. The data in the last segment of data with small error bars was obtained by Thomas Balonek at Foggy Bottom Observatory and Richard Miller's group at Georgia State University.

Figure 4. The approximate locations of observatories currently participating in the WEB Telescope.

object 0716+714. If this correlation is genuine, then the radio IDV cannot be due to refractive scintillation by the Galactic interstellar medium. This would constrain the radio emission region to be very compact, (apparent brightness temperatures of $\sim 10^{18}$ K) and would require an extremely high bulk velocity for the jet material — bulk Lorentz factors of more than 100, (Wagner 1996). Confirmation of radio IDV is a potentially revolutionary finding in the blazar field. The WEBT telescope combined with multi-longitude radio observations (or observation of a high declination blazar with with a single radio telescope at high latitude — since the radio observations can be done day and night) would provide much improved data for this study.

We seek to recruit WEBT participants from a diversity of locations. This permits multi-longitude observation and provides redundancy against inclement weather. We expect that the apertures generally available for this use, 0.3 – 2 meters, and the quality of the sites and CCD cameras will limit the WEBT to $\sim$19th magnitude. Most blazars of interest are brighter than this.

For WEBT observations, differential photometry using CCDs is performed. It is fortuitous that this accommodates a diversity of CCD camera types without loss of precision. First, reduced frames are produced. This requires correcting defects, subtracting estimates of read-bias and dark-current, and applying a flat-field correction. The reduced frames are a data product which are made available to all WEBT participants in FITS format to allow for full calibration of WEBT data.

Photometry is done on reduced frames to obtain instrumental magnitudes. In addition to the monitored blazar, instrumental magnitudes are also derived for 5–10 comparison stars. The Image Reduction and Analysis Facility (IRAF) is our primary means to obtain instrumental magnitudes. For isolated objects, the IRAF/daophot/phot routine is used. For monitoring the blazar flux at the core of an extended galaxy, we are developing a program for likelihood estimation of a PSF fit to the blazar while simultaneously fitting a detailed model of the galactic emission (for the observed band) convolved with the PSF. For each CCD frame, an estimate of the blazar standard magnitude and error are generated. The color response of the filter and CCD is used for this in conjunction with magnitudes of the reference stars in the appropriate filter bands.

We use the Internet to coordinate our efforts to obtain good data without wasting effort. Rapid communications allow us to respond appropriately to changes in the emission of the source, or to the status of the observations. As the Earth rotates beneath the source, observation is "handed off" from one observatory to another located to the west by e-mail. The "hand off" consists primarily of ascertaining that the new observatory is obtaining good data before discontinuing. The overlap in observations times provides for calibration. A real-time comparison of flux density during this overlap might eventually be possible.

A mechanism is in place to make WEBT communications available to all interested parties without distributing copious amounts of e-mail. E-mail sent to Blazar@gamma.bu.edu is posted on the Internet within 60 seconds at ftp://gamma.bu.edu/pub/mattox/blazar/new_mail. Since observers often don't have access to a Web browser, this file may also be obtained by anonymous FTP with a text-only terminal.

2.1. The Anticipated Evolution of the WEB Telescope

The WEBT observation of blazars would be of crucial value if we could expect the EGRET γ-ray observatory to continue to be as efficient as it was in 1991. Unfortunately, the EGRET spark-chamber gas is nearly depleted, so EGRET can now make only very limited observations. However, because of the expected improvements in the sensitivity of the Čerenkov γ-ray observatories, and the growing capabilities for x-ray observation through the launch of XMM, Astro-E, and AXAF, we expect that the WEB telescope capability of obtaining continuous, high-temporal-density, optical data will be very valuable. The only other means to obtain such data is with optical telescopes in space. XMM and the MeV γ-ray observatory INTEGRAL will carry optical monitors, but will not meet the full demand. And sufficient HST time is not available.

We currently activate the WEBT during multiwavelength campaigns when it is possible that the resultant optical data will be of high value. Because of the tremendous effort required, the WEBT now operates at a low duty cycle. We expect that the duty cycle for multi-longitude blazar observations can increase substantially if the brunt of such optical monitoring can eventually be born by automatic telescopes, such as those contemplated by GNAT. NASA now plans to launch a follow-on GeV γ-ray mission called GLAST in ~2004. It is anticipated that GLAST will simultaneously monitor the entire sky. Thus there will be a need for simultaneous optical monitoring of up to ~10 flaring blazars. Automatic optical telescopes will be crucial to the monitoring required during the GLAST mission.

Although the motivation for the organization of the WEB Telescope is to obtain data to understand the nature of the beamed continuum emission of blazars through the study of cross-band correlated variability, it is expected that a network of automatic optical telescopes could also be useful for studying other phenomena; especially transient phenomena, e.g.: binary stellar systems, quasar/galaxy lensing systems, microlensing events, and asteroseismology. The investigation of asteroseismology was the motivation for the formation of the Whole Earth Telescope (WET) in the 1980s, see http://wet.iitap.iastate.edu/wet/. The WET initially used PMTs, but is now using CCD cameras to improve sensitivity. We are in communication with WET participants about potential synergism between our efforts.

It is also likely that a network of automatic telescopes would be a very valuable asset for science education. The observational facilities, scientific agenda, images, and results could be made portrayed on a carefully developed Website which accommodates the expected wide dynamic range of scientific training of visitors. Also, the observational facilities could sometimes be available for remote use by educators and students.

References

Antonucci, R. 1993, ARA&A, 31, 473

Bloom, S.D. & Marscher, A.P. 1993, CGRO Sym, AIP Conf. Proc. #280, p. 578

Bloom, S.D., et al. 1997, ApJL, 490, 145L

Dermer, C. D., Schlickeiser, R., & Mastichiadis, A. 1992, A&A 256, L27

Hartman, R.C., et al. 1996, ApJ, 461, 698

Mattox, J.R., et al. 1997a, ApJ, 481, 95

Mattox, J.R., et al. 1997b, ApJ, 476, 692

Quirrenbach, A., et al. 1991 ApJ, 372, L71

Sikora, M., Begelman, M. C., & Rees, M. J. 1994 ApJ 421, 153

Takahashi, T., et al. 1996, ApJ, 470, L89

Tanihata et al. 98, in preparation

Wagner, S.J., et al. 1996, AJ 111, 2187

CCD Photometry with the A. R. Cross Telescope of the Rothney Astrophysical Observatory

E. F. Milone and P. Langill

RAO, University of Calgary, 2500 University Dr., NW, Calgary, AB, T2N 1N4 Canada; email: milone@acs.ucalgary.ca; plpl@iras.ucalgary.ca

Abstract. A *Photometrics Inc.* CCD camera and controller have been installed at the RAO on the ARCT, a 1.8-m telescope on an alt-alt mounting. The telescope was planned before the age of imaging, and differential rotation limits the lengths of exposures, and sometimes requires camera rotation. Photometric targets planned include variable stars in clusters, planetary nebulae, and solar system objects. A selection of recent observations illustrates the system's capability.

1. The Facility and Instrumentation

The ARCT has been in operation for the past three years since the replacement of its 1.5-m metal mirror with a 1.8-m Angel honeycomb-mirror (Milone and Clark 1990; http://www.acs.ucalgary.ca/~milone/rao.html, and linked documents). It is located in a 13-m dome at the RAO site in the foothills of the Canadian Rockies, ~ 75 km from the university campus in Calgary. The instrumentation is attached to a bonnette mounted to the telescope frame independently of the support structure for the 1.8-honeycomb mirror; three instruments can be accommodated. A Richardson-designed spectrograph is located at a mirror-fed East port and an infrared chopping system at the West port. The RAO has two CCD cameras, one of which is normally used on the Newtonian focus of the 41-cm telescope. A PHOTOMETRICS CH260 camera with a two-liter LN_2 dewar is mounted at the through-port on a rotatable frame which can be used to follow the field rotation that accompanies tracking of a paraxial target on the ARCT's alt-alt mounting. We find, however, that with the telescope guidance camera and star tracker, differential rotation permits up to 45s exposures without substantial smearing above that due to seeing.

The camera is operated at the $f/14$ Cassegrain focus which has an image scale of 8 arc-sec/mm. The chip is a thinned, back-illuminated, 1024^2 SITE chip with $25\mu m$ pixels, giving a chip scale of 0.2 arc-sec/pixel. Other properties are Gain = 4.6 e^-/adu and R = $5.9e^-$ rms.

The camera output is sent to a 686 class PC. A V FOR WINDOWS software package permits image examination and limited image analysis for checking purposes.

2. Data Reduction and Processing

The data are transferred to a Syquest 230 removable hard drive at the end of the evening and brought to the Astronomy Data Reduction Laboratory on the university campus, to a 233-MHz Pentium Pro computer (*jupiter*), running both LINUX and MICROSOFT NT. The data are periodically archived on 1.0 GB Syquest drive cartridges. IRAF is supported on *jupiter*, and on two other computers of a 5-machine LAN: a SPARC-5 (*neptune*) and a SPARC-1 (*saturn*) workstations running *Sun OS*. Two other PCs (a GATEWAY 2000 P5-90, *uranus*, and a 486, *triton*) are equipped with X-window software. A LEXMARK *Optra R+* printer is shared among the LAN computers. The LAN is part of a larger departmental network, PHAS, through which monochromatic and colour Lex-Mark printers are accessible. Finally both networks are connected to University of Calgary IBM *RS6000* servers running AIX.

In addition to IRAF, MIRA software is being evaluated by PL for student and observatory use.

3. Programs

3.1. Variables in Clusters

As we have argued in other venues (*e.g.*, Milone, *et al.* 1996), an important path to fundamental properties of the stars involves the analysis of multiwavelength light curves of selected variable stars in star clusters, the analysis of which will contribute fundamentally to our knowledge of the properties of stars and the reasons for stellar instabilities; variable stars that are in clusters provide distances to the clusters while the clusters' ages help to define the evolutionary stages of the variable stars. Once the brightness and colours of the stars are obtained, analytical tools that were partially developed at the University of Calgary are used to obtain the fundamental stellar properties. Specific current targets are the open clusters NGC 6791, NGC 7209, and NGC 752. An advantage of the bonette arrangement on the ARCT is the potential to carry out optical and infrared photometry as well as spectroscopy on the same night. For light curve work alone, a larger number of passbands offers distinct advantages for light curve analyses (see Kallrath and Milone 1999 for detailed discussions), especially with regard to sorting out the radiative properties of the stars.

3.2. Planetary Nebulae

Both the central star (or stars) and the nebula are targets. Specific nebulae under current study are NGC 2346, with its sometime variable nucleus. This program also involves observation of proto-planetary nebulae, specifically Frosty Leo and several IRAS-identified targets. This PPN work is in long-term support of programs carried out at CFHT and other observatories by members of the department (*e.g.*, by Langill, Kwok, & Hrivnak 1994; Kwok, Hrivnak, & Langill 1993; Volk, Kwok, and Hrivnak 1992).

3.3. Asteroids & Comets

This program is new to Calgary, and is being developed in tandem with increased interest in the solar system. A new Planetary Sciences program jointly sponsored by the Physics & Astronomy and Geology & Geophysics departments is under development. Eventually it is hoped that the RAO 's Baker-Nunn camera can be developed as a patrol telescope for both NEO and variable star discovery and monitoring and that the discovered objects can then be studied in detail with the ARCT and 41-cm telescopes.

4. Sample CCD Images and Results

Sample images appear in Figures 1-5.

5. Future Developments

Components of an infrared imager (on permanent loan from *ARC*) and a 5-12 μm infrared spectrograph (jointly developed with UCSD and UMinn) are under development and are expected to be completed this summer as additional instruments to support the RAO's programs.

If plans are implemented, the Baker-Nunn camera will be equipped with a large format CCD permitting FoVs of at least 2°. Software and interfacing will be carried out by Prof. M. Smith's group in the U of C's Computer & Electrical Engineering Department.

For all telescopes, a long-term goal is increased automation, so that observations can be taken at frequent intervals, regardless of the accessibility of the observatory in mid-winter.

Acknowledgments. We are grateful to PHOTOMETRICS INC. for providing significant challenges in both software and hardware management, to Fred Babott for installation and maintenance, to the students of the senior astrophysics laboratory course, ASPH 507, for discovering the system's shortcomings, and to graduate student Mike Mazur for conducting a number of critical tests of the system and software. This is RAO Publication Series B, No. 21.

References

Kallrath, J. & Milone, E. F. 1999, in Modeling and Analysis of Eclipsing Binary Stars. (New York: Springer Verlag). *in press*

Kwok, S., Hrivnak, B. J., & Langill, P. 1993, ApJ, 408, 586

Langill, P., Kwok, S., & Hrivnak, B. J. 1994, PASP, 106, 736

Milone, E. F., & Clark, T. A. 1990, BAAS, 22, 974.

Milone, E.F., McVean, J. R., Wilson, W. J. F., Kallrath, J., Schiller, S. J., Stagg, C. R., Mateo, M., & Yan, L. 1996, CR, 322, Ser. IIb, 177

Volk, K., Kwok, S., & Langill, P. 1992, ApJ, 391, 285

Figure 1. A V-band log-intensity image of a portion of the field of M51, obtained with the *ARCT* on April 29. Four 45-sec exposures obtained consecutively have been coadded to produce this frame. Note the shift in overlap frames due to the image rotation. The range is 800 pixels corresponding to $\sim$ 2.7 arc-min. No camera rotation was used for this auto-guided exposure. North is up. Note the spiral arm structure emerging from the nuclear region.

CCD Photometry with the A. R. Cross Telescope 107

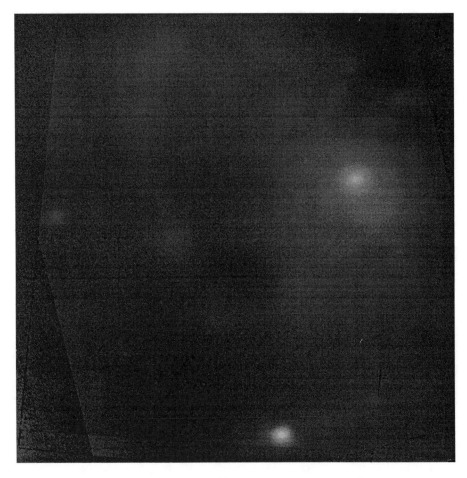

Figure 2. The same target as Fig. 1, but in the B-band.

Figure 3. The same target as Fig. 1, but in the R-band.

Figure 4. A log-intensity V-band image of M92, combining three co-added exposures taken in sequence on June 3, 1998.

Figure 5. A 50 sec I-band image of Frosty Leo taken Feb 2, 1998. The brighter southern lobe has an integrated brightness of $V = 11.90$ and (V-I)=2.65; corresponding values for the northern lobe are $V = 12.58$ and (V-I)=3.29 (as determined from CFHT observations).

A High Performance CCD System for Detection of Optical Supernova Remnants

T. Foster, D. Hube, J. Couch

Dept. of Physics, University of Alberta, Edmonton, Canada

B. Martin

The King's University College, Edmonton, Alberta, Canada

D. Routledge, F. Vaneldik

Dept. of Electrical Engineering, University of Alberta, Edmonton, Canada

Abstract. Many observatories, maintained by universities and colleges, are trading venerable photon counters for CCD detectors, extending the range of research small telescopes at these installations have performed in the past. The University of Alberta's Devon Astronomical Observatory has changed in this way, with the development of a wide field CCD camera system. The wide CCD frame eases photometric studies of stars, and permits new multicolor observations of low surface brightness, extended objects, such as supernova remnants. Exploiting all of the CCD's advantages means also building a system thoughtfully designed with the CCD's sensitivity in mind, allowing the observer to concentrate on data quality, rather than equipment, to enhance precision. We present a system constructed with the efficiency of the CCD in mind. It is hoped that other users of small telescopes equipped with CCDs will avoid the difficulties that we have experienced, and enjoy consistent, precise photometry.

1. Introduction

1.1. Overview of the Observatory

Located 30 km west of the city of Edmonton, Alberta, the Devon Astronomical Observatory is set in a beautiful wooded area near the University Botanic Gardens. The 0.5 m telescope has a f/3 primary mirror, and interchangeable front ends which provide f/8 and f/18 (Cassegrain and folded Cassegrain) foci, and a prime focus. Telescope and CCD system controls are operated by remote computer in a warm room adjacent to the dome.

The original 2 star photometer (at the f/8 folded Cassegrain focus) has been replaced with the addition of the CCD. A new front end was designed for holding a large format CCD camera at the f/3 prime focus, along with a corrector lens system to produce a coma free, $1°$ wide image plane (Figure 1). The camera requires a sensitive chip for the types of objects proposed for study (supernova remnants), as well as sharp images over the entire focal plane for stellar photometry. Focus and rotation mechanisms are built into the cylindrical camera housing. Finally, since multicolor photometry is common in our research repertoire, a filter holder capable of quick (1 - 2 seconds) position changes is necessary. Filters are carried in a bar slide.

After considering several chips and camera suppliers, we chose Spectra-Source's HPC-1 camera, equipped with an anti-reflection coated, thinned, back illuminated TK-512 sensor. At the time, this was the most sensitive device available that was within our budget and which would allow us to achieve our research goals. Since the f/3 primary cone of light is slightly narrowed by the corrector lenses to f/4; the 27 μm pixel size of the CCD produces a 23.6' wide field, with ~2.6"/pixel. Typical seeing at Devon is on the order of 3 - 5 arcseconds, so a stellar image may span a FWHM of 1.5 - 2 pixels. For stellar photometry, this is under-sampled, but for detecting faint extended objects, the sharp images are beneficial.

The system is designed for rapid, multicolor stellar photometry (variable stars and cluster work) and for detection/measurement of optical components of supernova remnants. Idendification of an optical SNR will be correlated to radio maps of the region, obtained by DR and FV. Since SNRs emit in discrete bands across a broad range of the spectrum, our filter sets include both UBVRI, and a broad selection of line filters.

Edmonton's prime observing seasons occur in the fall and the winter, where ambient nighttime temperatures range from 40 F to -22 F. Thus, extraordinary chip cooling methods, beyond a thermoelectric device are normally not needed to achieve a low dark current. The HPC-1 is equipped with a liquid chamber under the 2-stage Peltier device to allow for efficient heat removal by liquid circulation. This is especially necessary for our design, as the camera body is entirely enclosed within its housing. Antifreeze is circulated by a high capacity pump through a large reservoir for efficient heat exchange with the surroundings.

For an efficient observing run, the control software should be designed with rapid multicolor photometry in mind. To repeatedly image a star field in alternating colors in as small a time interval as possible, automation of filter changes between integrations is required. Software supplied with the HPC-1 is designed for simple imaging, so it was necessary to begin design and construction of custom control software. Although the cost was deemed very high, the camera function library files were purchased from the camera supplier. Development of the new camera control routine is well advanced.

1.2. From First Light to Today...

Construction of the mechanical components took place in the Dept. of Physics Machine Shop. As originally ordered, the chip was to have a new type of antireflection coat, forecast by Tektronix to be > 80% efficient in both the UV and in the visible. This coating was still under development at order time, and in the interim, SpectraSource supplied the camera with an engineering grade CCD chip, with the understanding that when the new coating became available, our specified chip would replace the lower grade device. In 1995, the camera was re-fit with a new chip.

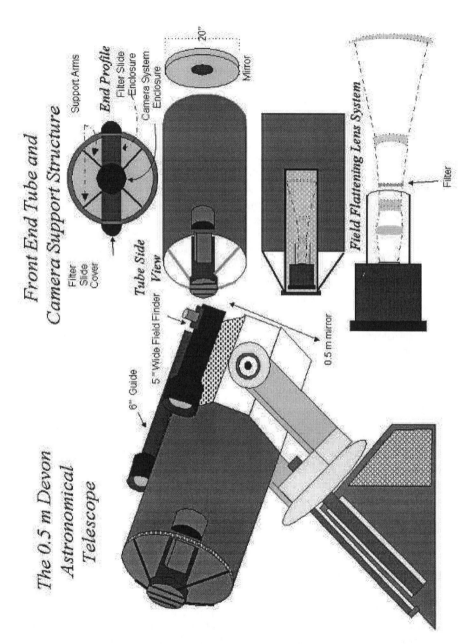

Figure 1. A drawing of the Devon telescope/CCD system, showing its principal features.

While development of the prime focus components proceeded, the camera itself was mounted at the telescope's f/8 Cassegrain focus for testing. This produced an 11' field, enough for some preliminary research. The old supernova remnant CTB-87, in Cygnus, was chosen as a candidate for detection, and some differential photometry was performed on the SX Phe type variable BL Cam. In each of these studies, the camera demonstrated its high sensitivity, as well as its ability for precision measurements, even under less than perfect photometric conditions.

The attempts to detect CTB-87 was the M.Sc. thesis project of graduate student Stefan Cartledge. Using B, V, OIII, and Hβ filters, images of the region were produced, and IRAF used for calibration and removal of all unsaturated star images. In the remaining image, non-stellar objects were fit with a polygonal aperture and photometry performed (Figure 2). Positive ID of them as pieces of CTB-87 remains inconclusive. Follow-up studies are planned.

The 13th magnitude star BL Camelopardalis (GD 428) was chosen for study due to its favorable location in Edmonton skies, and its short period of variability (0.0391 d.). Three nights of data, each a 3 - 4 hour run, were obtained under moderate to good photometric conditions with a V filter in place. Under these conditions, and in spite of an unidentified stray light source contaminating the field, the camera was able to provide 0.5% precision, yielding the light curve shown in Figure 3. We are confident that with more care taken in observing protocols and reductions, and care in eliminating stray light, our new system will achieve better than 0.1% precision.

In April of 1997, the assembly of the prime focus system was complete enough to allow installation on the telescope at Devon. Positive features of the system are:

- High quality of the optical system; sharp stellar images across the field.
- High sensitivity of the chip
- Wide image field; > 23'

In the previous configuration, the camera produced stars with an average FWHM of 4 - 5 pixels; star images well sampled enough for aperture photometry. However, the current prime focus arrangement produces stellar images with a FWHM of 1 - 2 pixels. While desirable for detection of faint supernova remnants, such sharpness must be traded for sampling when performing stellar measurements. This is accomplished by defocusing (see S. Howell, CCD Photometry Methodologies, in these proceedings).

2. Problems Encountered

2.1. Optics

Initial use of the prime focus system was not without problems. The foremost problem plaguing Devon has been a contamination of the CCD frame by a shadow-like image of the front-end tube profile. In most exposures, the frame would contain an uneven image of the tube and light path obstructions, which would normally be out of focus in any reflecting telescope design. The source of that feature was not immediately obvious and has been the subject of extensive debate.

A High Performance CCD System

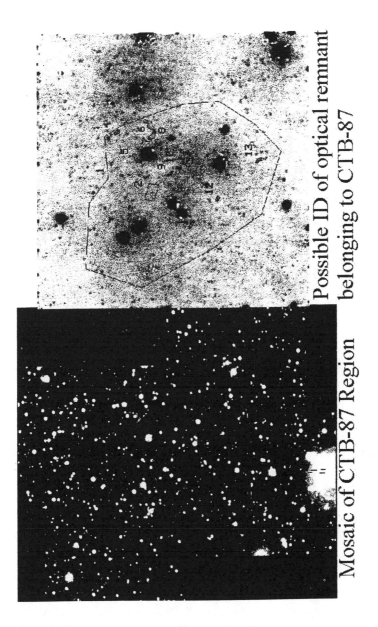

Figure 2. Results of preliminary study of the supernova remnant CTB-87 (α=20h 14m 10s, δ=+37°03'). The camera was still mounted at the f/8 Cassegrain focus for this project.

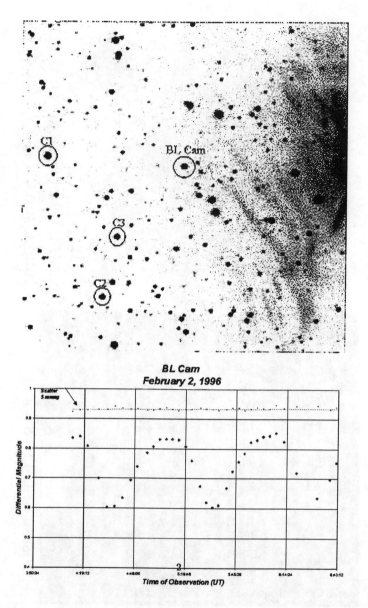

Figure 3. Light curve for the SX Phe type variable BL Cam (V filter). Note the precision (+/- 0.005 $m_{Inst.}$), even with stray light contamination Elimination of stray light can only help.

While the debate over the origin of the feature went unresolved, its serious impact on the precision of the Devon CCD system was unquestioned. The profile dominated frames taken in all but the darkest sky conditions (Figure 4). Flat fields of any kind (dome, twilight or night-sky) exhibited the feature, but their use in calibrating object images did not flatten the object field. Rather, the division of a flat field would result in a negative "ghost" left on the object frame. This uneven residual gradient seriously compromises aperture photometry, which relies on the consistency of the background.

It was agreed that the most robust test would be to place other detectors at the mount point of the camera to determine if they, too, show the same feature. This would eliminate the possibility of the feature originating from within the camera itself. Various detectors were placed at prime focus.

A simple look from prime focus down the tube showed several reflections. The various lenses present in the multi-element field corrector produced these reflected images. Presumably, the sensitive CCD is capable of seeing such images as well. Next, a 35mm camera body was mounted at prime and images of the dusk sky taken, with varying exposure lengths and camera positions. Close inspection of the resulting slides showed the same shadow image found in our CCD frames; the image was barely visible (probably due to ordinary film's relative insensitivity compared to a CCD) but there, nevertheless. Finally, a homemade CCD camera (Berry et al. CB-245 design) was obtained and installed. Again, images clearly showed the profile, albeit less prominently than with the HPC-1.

With these tests, it became clear that our woes were coming from one or more of the field corrector lenses. Apparently, the CCD is sensitive enough to image reflections from these elements, even though an antireflection coating (single layer MgF_2) is present on each surface.

2.2. Reflectivity of CCD Chips

Originally, our thoughts as to the cause of the feature concerned the CCD chip itself, and not the lens elements. Visual inspection of the chip showed its imaging surface to be highly reflective. We reasoned that this reflective surface might produce a pupil ghost, which in turn is imaged along with whatever is at focus. This led us to experiment with the camera, to determine whether we had been shipped a CCD with an AR coat, as originally ordered. Since the chip visually reflected so much light, we decided to quantitatively test this aspect. A spectrophotometer was enlisted to determine the reflectivity of the optical window/CCD chip combination across the UV-Vis-NIR spectrum. The results were surprising; 30-40% of the incoming signal is scattered/reflected away by the camera (most of this by the CCD itself). This indicated that we did not have what had been specified in the original order from SpectraSource. A visit to SpectraSource was arranged, to personally deliver the camera and have the chip inspected for evidence of an AR coating. It was determined that our chip was coated with something, although they did not know what. Fortunately, SITe (formerly Tektronix) still had records of what kind of chip was shipped to us. Apparently, the high Q.E. coating for UV and visible wavelengths (under development at the time of our order) never emerged. Instead, we were unknowingly shipped a CCD coated for enhanced UV applications, with a significant loss in efficiency at visible wavelengths. The UV enhancement (which gives about

Figure 4. The profile of the tube obstructions presented problems -it did not evenly flat field out, and it was sometimes strong enough to hinder identification of objects. Note the Cocoon Nebula (circled) hidden in the above image.

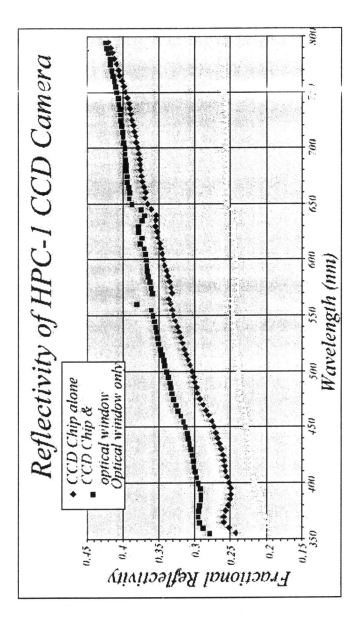

Figure 5. Reflectivity of the CCD chip/optical window composite, as seen by a spectrophotometer. Note the high reflectivity of the system (no visual antireflection coat), and the decreasing reflectivity towards the UV region, indicating a UV AR coat.

10% increase over the visual AR coat in Q.E. between 300 and 350 nm only) is of little value for astronomical applications, since the Earth's atmosphere is optically thick at 300nm and below. Even though our device does not meet the specifications for which we had contracted and paid, our camera manufacturer was unwilling to help us further.

2.3. Absorbed, Scattered and Stray Light

A problem which we had not seriously considered initially, and one that may have contributed to our reflection problems, is that of *stray light* in the optical system. In addition to contaminating the field, some of this light was originally part of the light cone: thus, signal strength is lost. Among the features in our design whose effects were not fully appreciated:

- The large number of lens elements in the optical path.
- The reflective anodized surfaces of the camera/filter housings
- The dusty primary mirror (a consequence of the local environment)
- The insufficiently baffled inner telescope tube walls

As is evident in the image of BL Cam and environs, and in figure 6, such improperly suppressed sources of stray light can do much to degrade a CCD frame, and to reduce accuracy achieved with aperture photometry.

In addition to external sources and internal reflections, one may encounter problems with internal sources as well. The dome housing our system is very dark, with no light sources (e.g. LEDs) from any equipment in proximity to the telescope. However, an unknown source consistently flooded the CCD frame, but only while using the I filter. This infrared source turned out to be the optical encoder LED used to indicate the position of the filter bar! Obviously, when designing a system, one must be mindful of sources not seen by the human eye, but easily seen by the CCD!

3. Solutions and Changes in Observing Habits

The goal in any observation where high precision is desired can be simply stated: obtain in the least amount of time, the highest S/N. One can think of the S/N as the contrast the CCD records an object having with its background. The equation for contrast in a CCD frame is simply:

$$C \equiv \frac{B(t) - Bo(t)}{Bo(t)} = \frac{B(t)}{Bo(t)} - 1$$

where B is defined as a linear surface brightness; e.g. DN/pixels^2. Bo is the background brightness, consisting of stray light, thermally generated noise, and anything else that shows up as unwanted charge within a pixel (cosmic rays, etc). The brightness on the CCD frame associated with each of these sources is time dependent. Contrast will generally depend on time: positively increasing and peaking at some point in the integration, and falling off as background (stray light) begins to dominate exposures (approaching the sky fog limit).

In obtaining precise data, the goal is to maximize C(t), in part accomplished by minimizing Bo(t). This can be done by:

- Choosing a darker site
- Combining several short exposures
- Cooling the chip to low temperatures

A High Performance CCD System

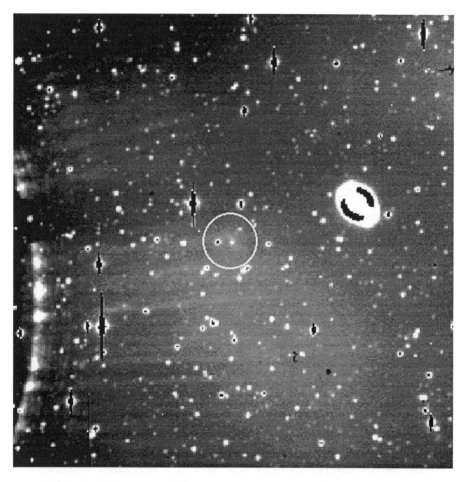

Figure 6. A portion of an M57 frame, overexposed to highlight the tiny spiral IC galaxy nearby. Note the scattered light in the field - probably the light of nearby Vega reflected from the shiny black anodized surface of the camera housing.

- Eliminating stray light in and around the system.

The first of these is impractical, unless designing a system literally from the ground up. The second is viable, but introduces an extra reduction step, which may add more noise than using a single, longer integration (R. Gilliland, 1991). Cooling is effective, but can reduce quantum efficiency, as explained by R. Tucker in these proceedings. If your chip is already cooled by a T.E. device, or other method, then more significant gains can be realized by addressing *stray light*.

3.1. Optics

For observations where high contrast is needed, having fewer lenses is better. This wisdom is truer for the CCD, which is more sensitive to low contrast situations than the eye in low light levels (the eye has a greater dynamic range). Even the highest quality lens with antireflection coatings still scatters $> 0.5\%$ of incoming light, and the absorption of light within the glass itself adds to this loss of signal. Particular to our system, the profile problem stems directly from lost light, reflected by each air-to-glass surface.

To alleviate this loss, we are currently testing the consequences of the complete removal of our corrector system, or its replacement by a simple, 2 element commercial coma corrector. Certainly, either of these two solutions is preferable to the other, namely, the addition of multi-coatings to every element in the corrector system. In particular, our design allows for the camera to remain sealed by the other lens elements with the optical window *removed*. Thus, an appropriate step for us is to remove this window, which contributes a significant amount to reflected (lost) signal (see figure 5). For those systems where removal is not feasible, ensure the optical window is tilted to direct reflected light away from the primary light path.

The ideas here seem contrary to common sense; removal of optics to correct an inherently optical problem, but they are sound, and should be considered by anyone looking hard at their system's sensitivity.

3.2. CCD Chips

Sources of signal loss are extended to include CCD chips themselves. A raw chip has a nasty reflectivity, contributing greatly to signal loss, and perhaps introducing spurious images. It is therefore wise to ensure that your chip is appropriate for your program, with the specifications you intended.

The wisdom imparted by many veteran CCD users involves really knowing one's camera. Many people spend hours at reduction time mapping bad pixels and hot pixels, building non-linearity curves, etc. So, why not know your chips reflectivity characteristics? Measure reflectivity of your chip across your operating spectrum. Knowing how much light you are losing at the chip, along with other aspects (size of telescope, quality of site, etc.) allows the observer to critically evaluate what kinds of observing can be done with one's system. This is an especially important determination to make in any newly CCD-equipped small telescope: our own observing agenda at Devon may have to be altered, with the new knowledge of our chip.

For those purchasing a system, we cannot stress how important it is to hear of others' experiences with particular companies. Do the research beforehand;

learn not only which company has quality products, but particularly good service/support as well. We would be pleased to offer anyone our full appraisal of our camera's manufacturer upon request.

3.3. Elimination of Stray Light

Significant gains in contrast can be realized by eliminating *stray light*. By their nature, mirrors scatter more light than lenses. Reflecting telescopes are subject to:
- External, off axis illumination (moonlight, bright sky conditions)
- Internal glancing reflections (tube walls, misaligned and/or dirty optics)

Both of these contribute to reduce contrast; i.e. reduced S/N for the CCD. One should therefore use clean optics, and ensure good mirror collimation. Consider enhanced aluminum coatings for an old mirror in need of a re-coat. The extra 5% reflectivity will prove valuable, especially for offsetting other sources of signal loss that little can be done about (e.g. a non AR coated chip). Attention must also be paid to the tube itself. Proper baffling includes tube extensions, well placed internal ribs, and a flat black finish on the inner surface and anything in the tube. We simply raised the existing structural ribs in our tube to an inner diameter of just less than 0.5m (our primary mirror diameter), and attached coarse sandpaper to all inner surfaces of the tube. The inside was then sprayed *flat* black, as were all surfaces of the camera housing. Finally, a tube extension was made from a lightweight thin foam mat, wrapped around the front end, and held by Velcro strips. The mat extends out well over the protruding camera housing, eliminating the possibility of dome reflected ambient light invading the tube.

To give one an idea of the effectiveness of such additions, consider Alan MacRobert's improved 12.5-inch (S&T, Dec. 1992, p. 696-99). MacRobert used a perhaps more effective internal blackening method; sawdust attached to the inner tube walls and soaked with ultra-flat black paint. With this and other improvements mentioned above, MacRobert saw a 0.6 visual magnitude gain in point source objects, over his original commercial reflector. This translates into a contrast gain of nearly two! To a CCD, a given low surface brightness object would be detectable in a shorter exposure; more free of thermal, cosmic ray, and other time dependent noise sources. For a sky fog limited site, the larger S/N is an important gain.

3.4. Observing Habits

Finally, the most important thing one can do to augment the above improvements is given as simple yet relevant advice by P. Massey and G. Jacoby (A.S.P. Conf. Ser., 23, 240). Visually assess every image as it comes off the telescope. Perhaps moonlight crept into the field slowly as the system tracked, or the aurora flared in the vicinity of your target, or the sky quality improved to the point where your stars now saturate. There is no more powerful way of 1) getting to know one's chip characteristics, and 2) eliminating data contaminated by stray light, than changing your observing habits and looking at every frame. Participate in your data collection!

4. Summary

Universities operating small telescopes not yet equipped with CCD technology often rapidly approach a CCD purchase and installation. It is important to remember that the replacement of the photocell with the CCD is a more involved operation than most believe, as our experience demonstrates. Existing telescopes, which work fine with photoelectric photometers, should have their structure and operating procedures critically reviewed prior to installation of a CCD detector. Extra care must be taken to completely eliminate stray light, whether it originates from internal reflections, or external sources. Observing agendas must be re-evaluated in view of the characteristics of your system, and observing habits changed to be more involved in critically looking at your data.

Acknowledgments. We are grateful to the organizers of this workshop, as well as all those who gave their advice and opinions concerning our system; particularly, A. Henden, R. Robb, W. Rosing, and E. Schmidt.

References

Clark, R. N. 1990, Visual Astronomy of the Deep Sky, Sky Publishing Corp., Cambridge MA

Gilliland, R. 1991, Details of Noise and Reduction Processes, A.S.P. Conference Series, S. B. Howell, 23, 69

MacRobert, A. 1992, S&T, 84, 696

Massey, P., Jacoby, G. 1991, CCD Data: The Good, the Bad, and the Ugly, A.S.P. Conference Series, S. B. Howell, 23, 240

Photometry at the Robotic Lunar Observatory in Flagstaff

James M Anderson
*Northern Arizona University/United States Geological Survey,
2255 N Gemini Dr, Flagstaff, AZ, 86001, USA*

Abstract. The Robotic Lunar Observatory is a project designed to produce radiometric images of the Moon for Earth-imaging spacecraft calibration over the wavelength range 350–2500 nm. Standard star observations provide information for instrument calibration and atmospheric extinction corrections. The project is currently developing data reduction software to process data taken by the automated observatory instrumentation. Several problems which limit the accuracy and precision of the observations are discussed.

1. Introduction

The Robotic Lunar Observatory (ROLO) is a telescope project dedicated to the radiometry of the Moon. The project is designed to provide absolute calibration images of the Moon for Earth-imaging spacecraft and is sponsored by the Earth Observing System, a part of NASA's Earth Science Enterprise (formerly the Mission to Planet Earth program). Project goals include the production of a photometric model of the lunar surface for approximately 120 000 selenographic grid points on the Moon which will allow radiance models of the Moon to be produced with errors under 1% relative and 2.5% absolute (Kieffer & Wildey 1996).

2. The Observatory

The ROLO observatory is located atop McMillan Mesa at the Flagstaff Field Center of the US Geological Survey in Flagstaff, Arizona. The project uses a standard fork mount designed for a 41 cm diameter telescope manufactured by DFM Engineering Co. in Longmont, Colorado. Attached to this mount are a visible/near-infrared (VNIR) camera and a short-wave infrared (SWIR) camera. The VNIR detector is a Thomson 7895B CCD purchased from Photometrics Ltd. which has 512×512 square pixels. The SWIR instrument was built by the infrared group at Steward Observatory and uses a 256×256 square pixel HgCdTe PICNIC array supplied by Rockwell International. Both cameras use narrowband interference filters for wavelength selection; VNIR uses 23 filters covering 350–950 nm and SWIR uses 9 filters covering 950–2500 nm. Attached to each instrument is a 21 cm diameter telescope with the SWIR telescope boresighted to the VNIR telescope. The focal lengths of both telescopes are near one meter, yielding angular sizes per pixel of $4.0''$ for VNIR and $8.2''$ for SWIR.

Neutral density filters are used to attenuate the photon flux when imaging the Moon in order to avoid overexposing the detectors. (A 1.2 cm diameter aperture stop was used instead of the neutral density filters prior to 1997 October.)

ROLO attempts to observe every clear night during the two bright weeks of each month surrounding Full Moon. A telescope operator is required to make weather related decisions and to open and close the dome, but other observing operations are controlled by an automated control system. Four personal computers are used for real-time control of the observatory. One computer controls the telescope mount, separate computers control the two camera instruments, and the fourth computer acts as the master controller. Using an ephemeris generated nightly and a database of prior observations, the master computer makes observing decisions in real-time, computing optimum exposure times based on predicted object brightness and extinction values; a crude phase and wavelength dependent photometric model of the Moon is used to predict the instantaneous brightness of the Moon. Observations of the Moon are made at half-hour intervals to provide adequate phase coverage. ROLO additionally observes bright standard stars from an internal list of 190 stars, many of which were taken from the list of primary standard stars defining the *uvby* system (Crawford & Barnes 1970). The master computer selects several stars from this list based upon the time of year and these stars are observed repeatedly throughout the night to determine a nightly extinction. Observations of the remaining standard stars fill the remainder of the night to provide absolute calibration of the instrumentation. The observing sequence is to point to an object, take images through the entire set of filters, and then go on to the next object.

The VNIR instrument started regular observations in late 1995 and the SWIR camera was added in 1997 October. The data collected thus far by ROLO during bright runs are in excess of 50 GB and fill 94 CD-ROMs. The data contain over 400 000 images; with the addition of SWIR, ROLO is currently taking images at a rate of $\sim$ 250 000 images per year. ROLO personnel are presently in the process of writing automated processing software to reduce these data with minimal human interaction. This reduction software uses the Integrated Software for Imaging Spectrometers (ISIS) software package developed by the Astrogeology Program of the US Geological Survey (Gaddis et al. 1997) for file and data structure handling but all other aspects of data processing are being specially written to accommodate the ROLO data. In order to meet the ROLO error budget goals, many problems must be overcome by proper instrument design, observing procedure, and software processing. Some of the difficulties encountered within the ROLO project are discussed in the following section.

3. Difficulties

3.1. Dark Current and Bias Levels

Because ROLO Moon and calibration (flat field) images are taken near the limit of the linear full well depth, multi-pinned-phase (MPP) mode operation of the VNIR CCD is not utilized to prevent charge spreading. This results in a relatively high dark current which must be removed from images. SWIR, although having a small dark current rate, has a highly non-linear bias level response as a function of exposure time. Being sensitive to longer wavelengths, SWIR data

Figure 1.   RFI Contaminated Image.

also contain a significant thermal response to warm objects such as the telescope. Careful determinations of the background levels are made at the beginning and end of each night. Temperature measurements are made throughout the night to enable compensation for instrumental changes by data reduction software using instrument response trends determined during instrument characterization tests.

3.2. RFI

Approximately 50 meters to the southwest of the ROLO observatory is an antenna used by the local fire and police departments for digital communications. Radio frequency interference (RFI) due to this antenna has caused serious problems for ROLO observations. Figure 1 shows an example of noise in a VNIR dark current image produced by this RFI. The light bands are caused by RFI induced bias level shifts in the A/D converter created during the readout of the array. These bands have average values up to 20 DN higher than the normal bias levels in addition to having greatly increased noise levels. Although instrumentation changes to more effectively electrically isolate the CCD equipment in 1996 August greatly reduced this problem, many thousands of otherwise useful images were contaminated by this effect. Software was written to detect this band shift in raw data and remove it to the best degree possible. However, the start and stop locations of the RFI normally occur in the middle of a row readout and determining the exact locations of these changes is difficult in complex images. It is also not possible to remove the additional noise corresponding to the shifts.

3.3. Focus

The full aperture configuration of VNIR suffered from poorly figured optics for many years until the entire VNIR telescope was replaced in late 1997. Figure 2 shows an example of an optimum stellar focus prior to the optics replacement. The full width at half maximum (FWHM) is roughly 8 pixels — more than 30". In addition, the VNIR and SWIR instruments are not capable of being adjusted for focus changes throughout the night. Stellar images therefore have complicated profiles and vary significantly in shape throughout the night, frequently consisting of out of focus doughnuts. Because the point spread function contains significant power at distances of more than one arcminute from the stellar center (King 1971), accurate CCD aperture photometry is non-trivial for ROLO data. The fraction of the total stellar energy which falls within a fixed sized aperture varies as the focus changes. Stellar integration software which changes the aperture size based on a measure of the focus for each image was estimated to be highly susceptible to systematic effects. Instead, ROLO software uses a fixed aperture size chosen to be large enough that the fractional change in the integrated stellar energy are small for expected focus changes, yet small enough that noise in the background level does not dominate the measurement process. In 1997 October the original VNIR telescope was entirely replaced when the SWIR instrument was added to the mount. New primary and secondary mirrors fixed the VNIR optics problems. In addition, both the VNIR and SWIR telescope tubes were constructed of a carbon composite material to reduce thermal contraction and expansion as well as to reduce the equipment mass on the mount. This change significantly reduced the focus change throughout the night, but did not eliminate the problem. The data reduction software continues to use a relatively large integration aperture for the new optics configuration, although this aperture size is 40% smaller than is used for the original optical configuration and background subtraction introduces less uncertainty into the measurement process.

Although the out of focus images present problems for most of the project's stellar observations, they do have the advantage that more photons can be detected without exceeding the linear range of the instrument. Spreading the starlight over tens or hundreds of pixels allows many more photons to be detected in longer exposures. Since photon shot noise goes as the square root of the number of photons detected, the fractional uncertainty in the measurement is reduced. Multiple calibration observations of bright stars by ROLO frequently have residual standard deviations of less than 0.005 magnitudes.

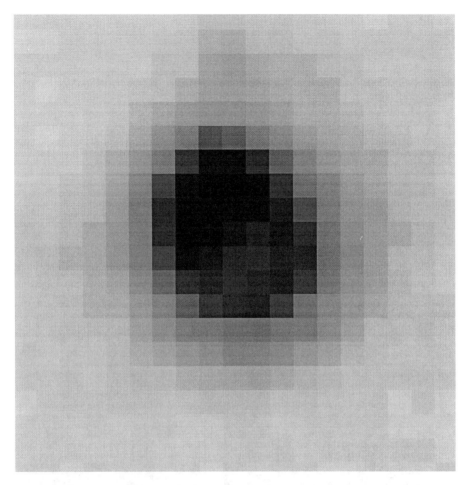

Figure 2. VNIR Focus: A stellar image is shown for an optimum focus position during 1997 March. Each pixel is 4″ on a side.

3.4. Flat Fielding

The original VNIR instrument used a small 1.2 cm aperture to attenuate lunar observations. This aperture stop created a telescope with an effective focal ratio of nearly 80, eliminating the focus problems mentioned above but causing flat fielding to be problematic. Figure 3 shows an example of a typical flat field image for this small aperture telescope. Dust accumulated on the CCD camera window at a rate of several specks per night. Since the dust obscuration changes the flat field value by 50% or more over a 5×5 pixel area, significant problems are encountered trying to properly flat field lunar images with only single calibration images per night. The variations in the lunar surface due to maria, craters, rays, etc. make detecting small flat field changes difficult. The filter movement between images needs to reposition the filters to within one pixel

(19 µm for VNIR) in order for dust specks to be positioned over the same detector positions. Minor position shifts or dust speck changes often result in 10% errors in the "corrected" images. Recognizing these problems, the small aperture was replaced by a neutral density filter in 1997 October to allow observations of the Moon to be made using the full telescope aperture. The additional ray paths present in the full aperture system spread the dust speck shadows over large areas of the CCD so that changes are smaller than 0.2% for any single CCD pixel. However, thousands of Moon images were taken with the small aperture system and need to be properly calibrated. At this time, no improvement over nightly flat field correction images has been implemented.

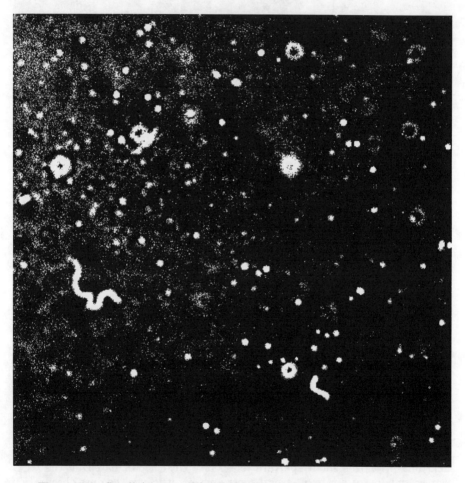

Figure 3. Small Aperture Flat Field: Small bright dots correspond to dust specks on the CCD window. Larger circular features correspond to dust specks on the interference filters. Long bright areas are probably cotton fibers remaining from cleaning efforts.

3.5. Sky Background

Because the ROLO project's main target is the Moon, ROLO naturally observes during the bright time of the month. Sky background levels are high even for the relatively bright standard stars observed by ROLO. In addition, the Moon is a bright source of position dependent scattered light in the atmosphere and the sky background frequently varies by 50% or more over the $8' \times 8'$ images taken for standard stars. Sky removal is thus a critical element of star processing. Several attempts at fitting a spatially variant sky background have so far proven to produce larger residuals than a standard sky annulus method. Two dimensional third and higher order Chebyshev polynomials were fit to the sky background pixels over two arcminutes away from the ROLO standard stars. The removal of the sky background using these fitted sky levels yielded residuals in the stellar summations which were more than 1.5 times larger than the residuals obtained by subtracting the average sky level in an annulus surrounding the stars. Small changes in the stellar profile from image to image are apparently able to affect the sky background fit even at large distances from the stellar centers.

In addition to producing difficulties with stellar measurements, the scattered moonlight also strongly influences measurements of the lunar flux. The amount of scattered light varies over the Moon so that removal procedures for proper calibration of the images must account for scattering from a non-uniform surface brightness object whose shape and orientation change dramatically with time. ROLO processing software does not yet adequately cope with this problem; only a constant sky brightness level is subtracted from the image. Future algorithms will include better light scattering models to remove the sky background from the Moon.

3.6. Non-Photometric Nights

In addition to the above problems, ROLO observes the Moon on nights which would not be considered photometric by normal astronomical standards. These observations are made because ROLO needs to obtain as much information as possible within a limited project lifetime to fill out the phase/libration space of observable lunar viewing geometry and produce a reliable photometric model. Using the observations of the standard stars, the reduction software currently computes a time dependent nightly extinction correction for each filter. Data processing for some nights during which the extinction has changed by as much as 40% seem to produce reliable results. In the future, ROLO personnel plan to develop software which will combine the data from all filters to produce an atmospheric composition model which is spatially and time variable in order to produce better extinction estimates.

4. Conclusion

ROLO processing software is still in its early stages of development. During 1998 April, all of the VNIR data were processed to produce exoatmospheric images of the Moon and extinction corrected irradiances for the standard stars. Although examinations of these data are still preliminary, some results from the stellar reduction can be given. Nightly residuals from producing exoatmospheric magnitudes of the standard stars have standard deviations of 0.03–0.02 mag for the original VNIR telescope and are usually less than 0.015 mag for the new VNIR instrument. However, a 5% scatter remains in the night to night comparisons of stellar brightnesses for the 2.5 years of data reduced. Averages of the stellar brightnesses are within 2% of published *uvby* magnitudes even though no color corrections were made for extinction calculations or for conversion to the standard *uvby* bandpass system. It is believed that the remaining scatter in the data are related to instrument drifts and poor extinction measurements. With continued observations and software development it may indeed be possible for ROLO to meet its goal of under 1% relative errors for the Moon.

Acknowledgments. The United States Geological Survey part of ROLO has been supported by NASA contract S-41359-F. The Northern Arizona University part of ROLO has been supported by NASA grants NAG 5-1761 and NAG 5-2159 and NASA contract NAS 5-96084.

References

Crawford, D. L. & Barnes, J. V. 1970, AJ, 75 978

Gaddis, L., et al. 1997, Lunar and Planet. Sci. Conf., 28, 387

Kieffer, H. H., & Wildey, R. L. 1996, J. Atmospheric and Oceanic Technol., 13, 360

King, I. R. 1971, PASP, 83, 199

GNAT Engineering Issues: Thermal Stability of a Prototype Imaging Photometer

Patrick R. Craine

School of Engineering, U.C. Berkeley, Berkeley, CA 94709 USA

Abstract. The Global Network of Astronomical Telescopes (GNAT) is in the process of evaluating a scale model prototype of the planned network of 1.0- meter photometric imaging telescopes. We report here on early experiences with the thermal stability of the GNAT prototype imaging photometer.

1. Introduction

The Global Network of Astronomical Telescopes (GNAT) has begun conducting a series of equipment evaluations to calibrate and determine the photometric precision attainable with a 0.5 meter version of a future GNAT telescope and a CCD imaging device. The initial goal of GNAT is to create a system of instruments that can be used as precise automated imaging photometers (Crawford and Craine, 1996). The imaging device being used at present is a PV-CCD camera with a TK512 detector chip. In order for this unit to create photometrically precise images it is necessary to keep all electronic noise to a minimum. The best way to do this is to keep the camera cryostat at a very cold, constant temperature (Buil, 1991). In this paper the thermal stability of the prototype GNAT cryostat and CCD detector will be detailed; assessment of photometric precision of the system is the subject of ongoing studies.

2. Instrumentation

GNAT is currently operating a PV-CCD imaging camera (Southwest Cryostatics, Tucson, AZ). This unit is equipped with a TK512 detector chip (SITe, Inc., Seattle, WA). The chip is cooled by a four stage thermoelectric cooler. To enable thermal stability, a vacuum must be maintained in the cryostat by periodic evacuation. Heat is removed from the thermoelectric cooler by a water sleeve which cycles through an ice bath and contacts a heat sink on the back of the CCD detector. To maintain a constant minimum chip temperature it is necessary to keep the cooling water reservoir near one degree centigrade.

The cryostat is evacuated with a two stage model 1405 Duoseal vacuum pump (Welch Vacuum Technology, Inc., Skokie, IL). A vacuum hose connects the pump to a length of 0.25 inch copper tubing which is run through a cold trap to precipitate foreign matter out of the air and prevent it from settling in the cryostat, potentially contaminating the CCD detector. The cold trap is a vacuum bottle which is filled with chipped dry ice. The copper tube connects the pump to the cryostat valve be means of a vinyl hose. More recently the cold trap has been replaced with a VisiTrap molecular sieve (MV Products, North Billerica, MA). This unit is a great convenience since it eliminates the step of charging the system with dry ice. All vacuum system junctions are coated with a high density Dow vacuum grease. The evacuation procedure involves pumping on the system for one half hour before opening the vacuum valve to the camera cryostat. We then pump the cryostat for approximately 10 to 12 hours. The vacuum valve on the cryostat is closed and the pump is shut down. The camera is then powered on and a cooling curve (chip temperature as a function of time) is immediately generated to confirm that the camera is quickly (about one half hour) cooling to the minimum temperature dictated by the current thermoelectric cooler thermostat setting. Attainment of this goal is taken as confirmation that a good evacuation of the cryostat was achieved.

A thermocouple attached to the CCD substrate inside the cryostat sends voltage information to a pair of electrical jacks external to the cryostat. The thermocouple is calibrated such that one millivolt corresponds to 0.1 Kelvins. Until recently all temperature data from the CCD was recorded manually using a Fluke Model 75 multi-meter (Fluke Mfg Co. Inc., Everett, WA), which can measure millivolts. A TES DMM-2730 recording digital multi-meter (JDR Microdevices, San Jose, CA), which can measure hundredths of a volt, has now been installed to record camera temperatures at set intervals through an RS-232 port to a Toshiba 486 laptop PC. The latter setup, although less sensitive, has the huge advantage of operating untended over long periods of time.

The camera is controlled by a low-end 486 PC running DOS based software written by the camera supplier. This software is used to set the camera integration time and to trigger the shutter. This unit can be slaved to the telescope control computer so that the camera can be operated in a fully automated mode.

3. Data Collection

The data used in this project have been collected by monitoring the voltage output of the thermocouple attached to the CCD (which corresponds to the temperature of the chip). The data recorded are the time since the CCD was turned on, the temperature of the CCD chip and the ambient temperature of the observatory. Before power is delivered to the CCD ice must be added to the reservoir through which the thermoelectric cooler's water sleeve runs. This bath should reach a temperature near one degree centigrade. At intervals of approximately five minutes the voltage was recorded. Recently this has been done with a digital recording multi-meter and a portable computer.

A typical cooling curve made several weeks after an evacuation is shown in Figure 1. Temperature was recorded approximately every five minutes. This figure indicates the minimum temperature, the time to achieve minimum temperature, and the thermal stability of the cooled detector. In this case we see the effects of cooling the chip in a poorly evacuated cryostat.

Figure 2 is identical to Figure 1, except that the cryostat has been recently evacuated. The chip is now operating at a much lower temperature with much greater stability.

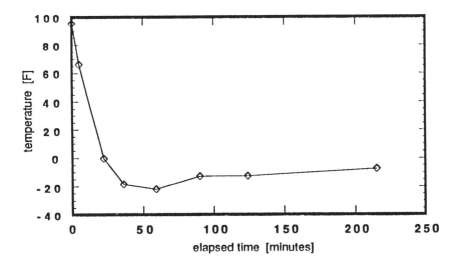

Figure 1. PV-CCD camera cool down curve for a poorly evacuated cryostat. Note that the minimum temperature is about -23F and the temperature appears poorly regulated.

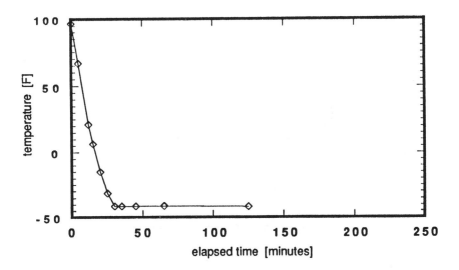

Figure 2. PV-CCD camera cool down curve for a newly evacuated cryostat (10.5 hours pump time). Note that the minimum temperature is about -45F and the temperature is well regulated.

Figure 3 shows the temperature of the CCD over a 22 hour interval. The temperature of the CCD was automatically recorded every 15 seconds. The cooling water was kept near one degree centigrade until approximately 780 minutes after the cool down was begun.

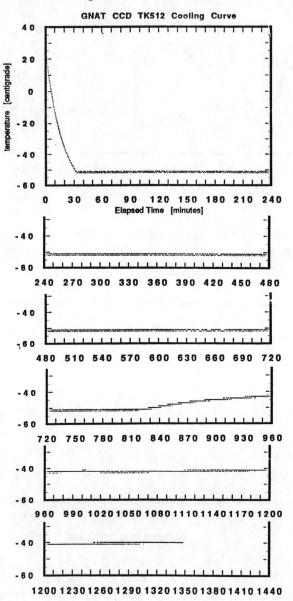

Figure 3. A PV-CCD camera temperature curve from power-on throughout the following 22 hours. See text for a discussion of this plot.

We also monitored data relating both minimum temperature and time required to reach minimum temperature after camera power-on as a function of days elapsed since cryostat evacuation. These data are shown in Figures 4 and 5.

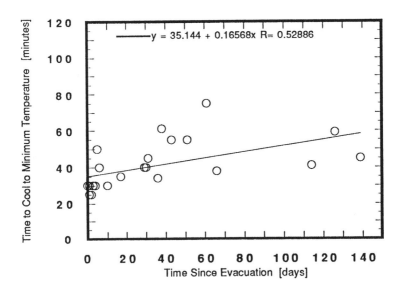

Figure 4. PV-CCD camera cooling time as a function of time since cryostat evacuation.

It became apparent that the original prototype cryostat was not performing optimally, in that it apparently had a slight leak giving rise to numerous problems of achieving and maintaining desirable temperatures. The camera was subsequently re-sealed and the detector was provided with additional insulation. Following these improvements cooling curves were again obtained. The temperature stability remained excellent, however this was achieved without resorting to icing the cooling water, but rather running it at ambient temperatures. We are aware, however, that with high summer temperatures in Tucson, the camera cools noticeably more slowly, and the minimum temperature can only be achieved by further cooling the water bath. A typical cooling curve for the new cryostat is shown in Figure 6.

We also note a diagnostic graph showing the effects of inadequate mechanical contact between the water circuit and the heat sink (Figure 7). This problem occurred during an instrument change. When the camera did not cool properly, the cooler was cycled and restarted, still without cooling properly. It was necessary to de-mount the camera and reassemble it to discover the problem.

Finally, a series of dark frames was made during the process of cooling the chip after camera power-on. These dark frames were each of 120 seconds duration, and the time elapsed since power-on was recorded for each.

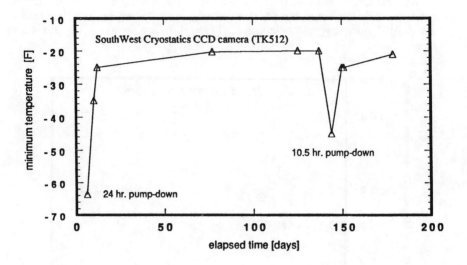

Figure 5. Minimum temperature achieved by the PV-CCD camera as a function of time. Two cryostat evacuation events are shown, one of 24 hours duration, the other of 10.5 hours duration.

4. Data Analysis

Figures 1 and 2 clearly show the gross effects of decay of slow decay of the vacuum in the camera cryostat. For the poorly evacuated cryostat, the cool down time is significantly longer, the minimum temperature is much higher, and thermal stability is somewhat poorer. The latter effect is a result of failure of the camera thermostat to regulate the temperature properly at the relatively high temperatures of a poorly evacuated camera. Clearly, the cryostat must be regularly monitored and sufficiently frequently evacuated to maintain the desired temperature.

In Figure 3 we take a closer look at the camera behavior during the course of a complete night. A twenty-two hour cool down was performed to determine the thermal stability of the chip over an extended period with careful cooling of the reservoir. This test was run shortly after an evacuation and a resetting of the thermoelectric cooler thermostat. Minimum temperature was achieved after about thirty minutes. The chip temperature was recorded automatically every 15 seconds in units of one degree centigrade (the resolution of the multimeter) which gives the data in Figure 3 the quantized appearance. After the first 780 minutes we stopped adding ice to the reservoir. During the time from 60 to 660 minutes the mean chip temperature was -51.5 centigrade with a standard deviation of 0.5 degree. From 810 - 960 minutes the mean temperature was -46.46 centigrade with a standard deviation of 2.78 degrees. From 960 minutes to the end of the test the mean temperature was - 41.37 Centigrade with a

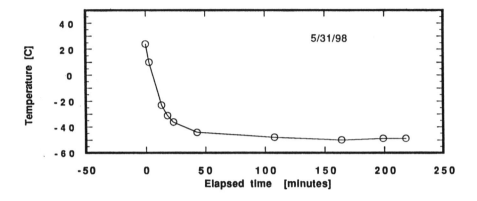

Figure 6. PV-CCD camera cool down curve following re-sealing of the cryostat and cryostat evacuation.

standard deviation of 0.856 degree. That is, after the last icing of the reservoir, the temperature of the chip steadily rose by about 10 centigrade to a fairly stable plateau.

We have also related the time to achieve minimum temperature to the number of days since evacuation. These data are graphed in Figure 4 for periods ranging between 0 and 140 days. These data points are fit by a least squares solution shown by the straight line. It is apparent that the time required to reach minimum temperature increases slowly as a function of time since cryostat evacuation. The camera typically takes between 30 and 50 minutes to reach operating temperature. Occasionally the camera took an anomalously long time to reach operating temperature, which seems to have been caused by particularly high ambient temperatures at the beginning of the cooling cycle.

The minimum temperature achievable can also be represented as a function of time since evacuation. An experimental sample cycle is shown in Figure 5. The minimum temperature increases with time since pump-down very quickly after the first week, and then reaches a plateau. These data are for the original cryostat, and suggest that a pump-down time of 12 or more hours is desirable. With a more secure seal the pumping frequency goes down to every couple of weeks.

Figure 6 simply verifies that after re-sealing, the cryostat was able to perform optimally without a stringent requirement for maintaining the cooling bath at a very low temperature. However, experience has shown that a high ambient temperature (100F) can still overwhelm the cooling system, and auxiliary cooling of the water is required. The data in Figure 7 serve as a reminder that even

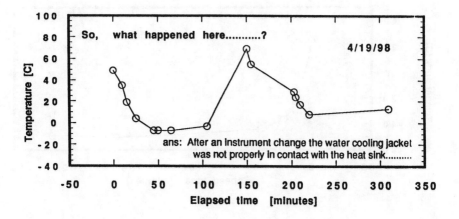

Figure 7. Similar to Figure 6, except taken following a reinstallation of the camera in which a slight wedge was introduced between the heat sink and the cooling coil.

simple things can go wrong: in this case the end of a string hose tie lodged under the heat transfer block providing a very slight wedge which was still enough to wreak havoc with the chip cooling.

Finally, in Figure 8 we present data to show the importance of these efforts to the photometric measurement process. These data were derived from the 120 second dark frames made during a camera cool down cycle. For each dark frame, five 50-pixel square regions were selected across the image and the standard deviation of the mean pixel intensity was calculated. These five values were averaged and taken as a measure of the noise in the image, called Noise1 in Figure 8. From a noise standpoint, it is clear that the thermal stability we have been able to achieve is completely adequate to enable the best possible photometry with this system. Further, the operating temperatures we routinely achieve with the thermoelectric cooler stack are well onto the flat part of this curve, suggesting that for our system we will gain little advantage from cooling further.

5. Conclusions

These data allow us to reach several conclusions:

1. With appropriate cooling of the water reservoir the CCD chip can be operated at temperatures stable to about 0.1-0.5 degree centigrade over extended periods of time.

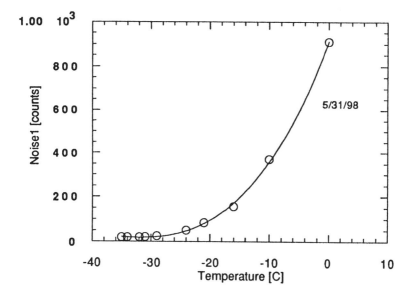

Figure 8. Dark frame noise as a function of temperature of the CCD chip in the PV-CCD camera.

2. The camera can easily be operated to at least -51 centigrade with proper operation. Given the speed at which this temperature was reached it may be possible to operate with the thermostat set even lower. This has not yet been attempted.

3. Periodic evacuation of the cryostat is essential to obtaining optimum performance of the camera. Determination of the need to evacuate the cryostat can best be made by monitoring the temperature behavior of the camera.

4. A securely sealed cryostat with a well insulated detector generally relieves the need to maintain a very cold water reservoir; however, during hot weather it is clear that a regulated water chiller will be necessary. It will be necessary for automated operation of the system to incorporate a regulated refrigeration unit to cool the water reservoir, since icing the reservoir requires frequent human intervention.

5. Thermal noise is a strong function of the temperature of the detector and great care must be taken to maintain low and stable temperatures in order to attain photometric precision.

6. Acknowledgments

The author expresses appreciation to the staff of GNAT for access to materials and equipment, as well as personal guidance, during the course of a GNAT internship. He would also like to thank Dr. Lois J. Blondeau for support of the GNAT independent studies project.

References

Buil, C. 1991. CCD Astronomy (Richmond, Virginia: Willmann-Bell), p.125.
Crawford, D.L. and Craine, E.R. 1996. JRASC, Vol. 90, No. 4, p.224.

A 1.0 meter Automatic Imaging Telescope for CCD Photometry

James D. Wray

SciTech Astronomical Research, 21200 Todd Valley Road, Foresthill, CA 95631

Abstract. With the Global Network of Astronomical Telescopes (GNAT) as a stimulus, we have undertaken to develop a professional, fully automated telescope designed especially for precision CCD photometry. The goal is to produce a turn-key system which can be replicated at low cost and can be easily scaled to larger apertures. The prototype system has already been delivered to GNAT, and much of the testing of the prototype has led to improvements to the full-scale 1.0-m telescope described here.

1. Introduction

A SciTech Astronomical Research design for an Automatic Imaging Telescope (AIT) has been built and tested as a half scale prototype of a 1.0-meter version. (see discussion by E. Craine elsewhere in this volume). The prototype 0.5-m telescope has undergone extensive development testing at its GNAT Tucson site, and is currently supporting development of GNAT research and science education outreach programs.

Our own research objectives in this telescope development program comprise the development of cost-effective technologies targeted at bringing small telescope performance into the sub-arcsecond region in pointing, tracking and imaging under remote automated control. Our application objective is to use these results in developing an exceptional high performance AIT for networked remote scientific data acquisition especially well suited for efficient area imaging precision photometric measurement applications, all in an affordable turn-key package.

The present development testing program with the 0.5-m system has verified, foremost, the basic design concept of the STAR Class Telescope, demonstrating dynamic performance properties characteristic of appreciably larger, more massive telescopes. In a second area, feasibility demonstration for real-time mount model pointing and tracking error compensation, a relatively simple algorithm was shown effective in improving real-time pointing and tracking by nearly two orders of magnitude over the uncorrected mechanical mount system. The mechanical mount system was intentionally left in its raw fabricated state to provide the basis for this research experiment. In its design machined state, operating with the real-time mount model pointing and tracking corrections active, unguided tracking performance at the sub-arcsecond level appears likely over extended periods of time for the next generation telescope. Similarly, direct pointing accuracy in the arcsecond rms range has now become a target goal for the enhanced follow-on STAR Class Telescopes in the 1.0-m aperture class.

All design enhancements resulting from the prototype evaluation are presently being incorporated in an intermediate sized 0.8-m aperture telescope. When completed, this proof of concept telescope will provide performance demonstration/verification for a system effectively identical to that of the 1.0-m STAR Class AIT.

Features of the Enhanced STAR Class 1.0-m AIT:

Design

- Modified classical polar yoke equatorial mount
- Field-flattened optimized Ritchey-Chretien optical system
- Sub-arcsecond backlash and periodic error hardware constraints
- Mechanically stiff and solid for wind exposure with roll-off enclosures
- Dimensionally compact with short swing radii for minimal dome size

Features

- Stainless Steel triple-reduction roller drive systems
- All steel rigid shell structure
- Broad-based kinematically supported elements
- Invar thermal compensating focus metering truss
- Large effective-radius wrap-around precision encoder scales
- Compumotor brushless servo step-and-direction drive motors
- Fully integrated automatic autoguider and programmable filterwheel
- Autonomous fully automatic remote operation ATIS control system
- Continuous run-time mount model pointing-tracking error correction
- Low to null thermal strain optical elements
- Precision home index reference marks, +/-1 count, least significant bit

2. AIT Design Considerations and Enhancements

The 0.5-m telescope has served as an evaluation model for the GNAT AIT, both as a mechanical imaging system and as a platform for the software control system development; and as a means of enabling GNAT pilot programs such as establishing CCD imaging filter system standards and in providing research materials and data sets to participating institutions in a broad program supporting research and science education outreach. Thus the 0.5-m test program has been both multi-faceted and multi-disciplinary.

From these activities a number of issues arose which bear directly on the design of the AIT and which lead to specific enhancements for the design of the 1.0-m AIT. These issues included:

1. The effect of systematic micro-creep on mount model calibration stability

2. Minimizing concentricity errors in the mechanical system to reduce stress on the mount model computational system

3. Secondary mirror support stability under focus travel motion

4. All flexures be smooth and continuous (free from discontinuities) for repeatably accurate model compensation

5. Automatic focus maintenance requirement

6. Maximizing guide star area accessible by the offset guider system

7. Increase the number of filter positions according to the increase in the size of the instrument bay

Assessment of each of these issues produced conceptual and design changes which resulted in a number of enhancements in the 1.0-m AIT design as compared to the 0.5-m prototype. Each is described briefly in the following paragraphs.

1. Micro-creep in the roller-roller interface either requires a mount model that contains information on the absolute position of each individual roller together with individual calibrations for each roller, or

2. a roller concentricity precision for every roller sufficient to reduce the entire range of combined radial error to an amplitude smaller than the design performance value. Our approach to this follows the latter of the two, and our design angular error due to maximum combined radial runout is less than 1 arcsec, with the largest roller providing the largest individual contribution, 0.5 arcsec with a period of 1.8 hours. This design goal represents an improvement of a factor of 20 over the drive roller concentricity in the 0.5-m prototype.

3. Secondary mirror hysteresis (multi-valued kinematic function, i.e. flop) is avoided in the 1.0-m design with a two-stage focus travel secondary mirror support. The first stage travel is large, to accommodate adjustment

to different instrument packages. This coarse focus adjustment is manual and once obtained is locked into position with zero free-play. All components of a given instrument package are parfocalized, such that once an exact focus is obtained a perfectly operating thermal compensating focus metering truss support system would eliminate any further need for focus adjustment. This leaves both the requirement to obtain a precise initial fine focus and, due to the likely possibility that the focus metering system cannot exactly track all dynamic focus changes, any necessary further fine focus adjustments. Both of these fine focus adjustment requirements are met with the second stage focus adjustment system. The second stage focus adjustment acts through a small range constrained by the travel of dual radial sets of radially stiff, axially flexible stainless steel flex bands which rigidly constrain the radial motion of the secondary mirror cell while permitting small adjustments in the axial (focus controlling) direction. These adjustments are implemented in such a way that the firm support is always gravity loaded, and additional preload is applied down-tube to compensate the weakening gravity vector component with increasing zenith distance. The result is that the secondary mirror support system is rigidly constrained and, being designed such that it tracks Youngs Modulus within the continuous elastic range at all times, is therefore suitable for modeling as a single valued (predictably repeating) dynamical system.

4. The secondary focus control in the 0.5-m telescope is accomplished by a computer controlled motor. In the 1.0-m AIT focus stability will be maintained by a thermal compensating focus metering truss. This system, more than a simple invar support truss, incorporates an appropriate amount of opposing thermal strain from a high coefficient material (aluminum) to produce a null motion of the focal surface at the instrument package image plane. To do this all elements of the system are analyzed for their contribution to the focus metering budget, and the invar elements together with all the associated interface elements are considered as components in the focus metering truss system. The net result is a design effective null focus shift over a 20C ambient temperature range. Fine focus positioning by motor control is also available for any conditions which require it, and this control can be coupled to atmospheric effects on image quality for automatic focus control as necessary.

5. In a dedicated AIT system not only the secondary mirror flexure, but all flexures in the telescope need be smooth and continuous single-valued functions of all possible stress loadings encountered in telescope pointing and tracking. The same design paradigm applied to the secondary support system was also applied to the attachment of the invar metering truss to the telescope main-frame, the attachment of the primary mirror support structure to the metering truss and telescope main-frame, and the axial and radial restraint of the primary mirror in its support structure. Likewise, all bearing, adjustment and attachment assemblies throughout the telescope and integrated guider and instrument packages are preloaded to a design level higher than any possible gravitational release stress which may be encountered as a function of varying telescope pointing vector.

Thus the entire system design complies with the single-valued kinematic flexure model. This is a basic requirement necessary to insure that target objects are routinely positioned on the detector exactly as intended by the automatic operating system, a necessity for reliable and efficient automatic image data acquisition.

6. The prototype autoguider uses a rho - theta coordinate system where the theta position is established by positioning of a rotating stage and the rho position is constrained to the X coordinate range of the guider CCD detector array. Guide star catalog searches indicate that this concept necessitates incorporating rho stage motion as well to increase the radial coverage. This combination gives access to the entire area of the off-axis optical image. This change is being incorporated in the AIT design.

7. The filter wheel for the 1.0-m AIT carries twelve 50mm sq filters as compared to six filters in the 0.5-m telescope.

3. 1.0 meter AIT Design

The following enhancements have been built into the 1.0-m telescope design:

1.0-m design features

Table 1 Enhanced 1.0-m STAR Class AIT

```
Aperture              1.0 meters
Primary f/            2.2
Overall f/            8.0
EFL                   800 cm
FOV                   22 arc min sq on 50 mm sq detector
Field size            70 mm diagonal
Field Scale           25.8 arcsec / mm
Image Scale           38.8 micron / arcsec
Pixel Scale           0.62 arcsec / pixel   (@ 24 micron pixel)

Optical design        Optimized Ritchey Chretien
Optical material      ULE
Design PED*           80% energy in 24 micron square over
                          22 arc min dia field

Encoders              0.16 arcsec readout sample precision
Drive increment       0.025 arcsec
Max rate              0.1 radian/second per axis


Weight                4000 kg
Size                  See drawings

*    Point Energy Distribution
```

4. Discussion

Presently a 0.8-m version of this design is under construction scheduled for completion in mid-1999. Enhancements on the 0.5-m design intended for the 1.0-m telescope are being included in the 0.8-m telescope. This telescope will be used in a fully AIT mode, and its development at this time provides an ideal basis for performance demonstration and proof of concept for the enhanced STAR Class AIT as described in this paper.

Table 2 Enhanced 0.8-m STAR Class AIT

```
\begin{verbatim}
Aperture                0.8 meters
Primary f/              2.3
Overall f/              7.1
EFL                     577 cm
FOV                     0.5 deg sq on 50 mm sq detector
Field size              70 mm diagonal
Field Scale             35.7 arcsec / mm
Image Scale             28.0 micron / arcsec
Pixel Scale             0.86 arcsec / pixel   (@ 24 micron pixel)

Optical design          Optimized Ritchey Chretien
Optical material        ULE
Design PED*             80% energy in 24 micron square over
                           .5 deg dia field

Encoders                0.21 arcsec readout sample precision
Drive increment         0.025 arcsec
Max rate                0.1 radian/second per axis

Weight                  2000 kg
```

5. Conclusions

A 0.5-m prototype AIT has been used to refine the design concept for a follow-on 1.0-m AIT. This phase of the 0.5-m program, although continuing with the GNAT objectives of automated photometric imaging for multi-user science and educational applications, has already produced the intended interim result of providing design enhancements for the 1.0-m AIT in a number of significant areas. All of these enhancements are implemented in a 0.8-m AIT. With possible exception for design modifications arising from the 0.8 meter implementation the design of the 1.0-m AIT has been completely defined, allowing us to move forward with the hardware implementation of the enhanced 1.0-m STAR Class AIT fully optimized for the GNAT precision photometric measurement application.

A 1.0 meter Automatic Imaging Telescope 149

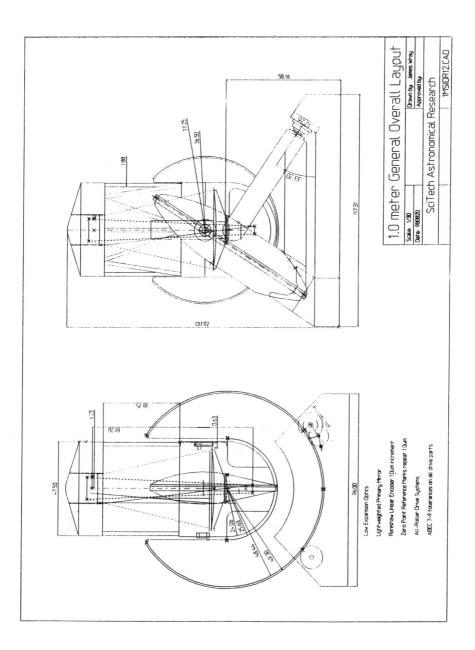

Figure 1. General layout drawings of the 1.0-m AIT.

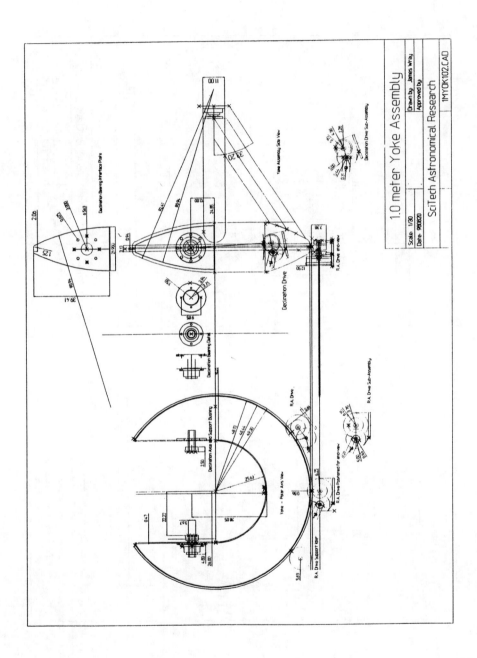

Figure 2. 1.0-m AIT RA yoke assembly layouts.

A 1.0 meter Automatic Imaging Telescope

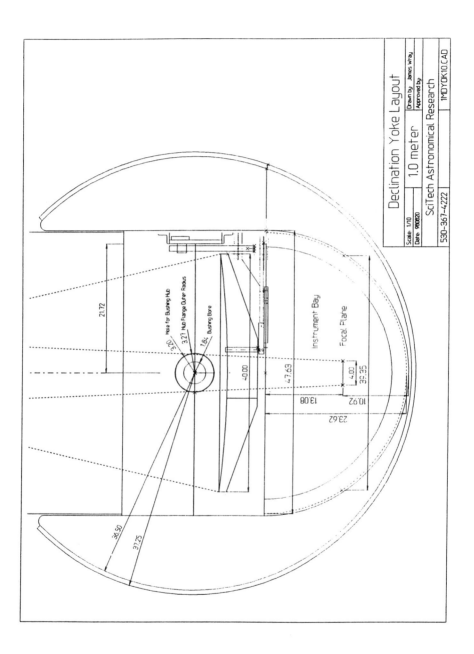

Figure 3. 1.0-m AIT declination yoke layout.

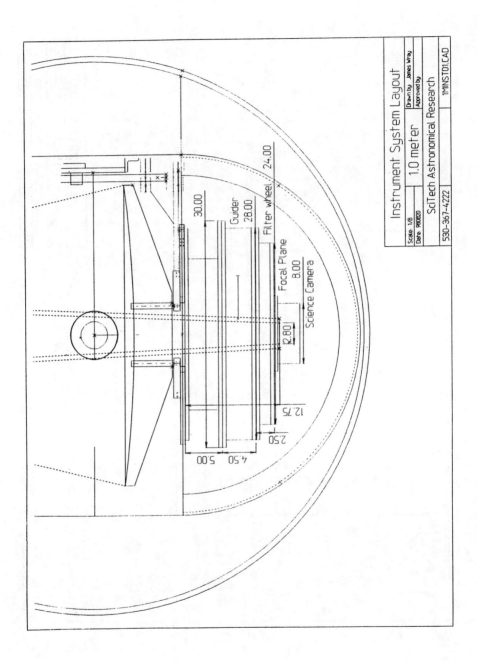

Figure 4.   1.0-m AIT instrument bay layout.

A 1.0 meter Automatic Imaging Telescope

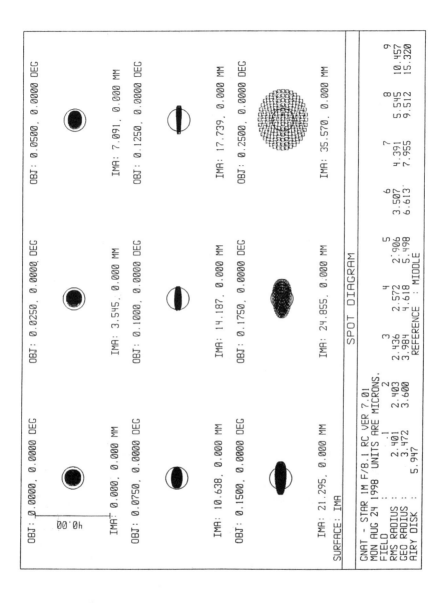

Figure 5. Spot diagrams for the focal plane energy distributions: 1.0-m AIT.

ARLO: Automated CCD Reductions for the Construction of an All-sky Catalog of Photometric Calibrators

R. Casalegno, B. Bucciarelli

Osservatorio Astronomico di Torino, Pino Torinese TO, Italy

J. Garcia, B. M. Lasker

Space Telescope Science Institute, Baltimore MD, USA

Abstract. ARLO is a fully-automated set of IRAF scripts dedicated to the photometric reduction of CCD data for the calibration of photographic sky surveys. Particular attention has been given to the tuning of critical IRAF image-processing parameters, as well as to photometric accuracy, database organization, quality assurance, and error analysis. As ARLO is likely to meeting other requirements for automated reductions of CCD photometry, it is being distributed from the OATo WWW site.

1. The Scientific Case

A few years ago the Space Telescope Science Institute (ST ScI) and the Osservatorio Astronomico di Torino (OATo) started a scientific collaboration to produce a new generation Guide Star Catalog (GSC–II, Lasker et al. 1995), based on first-, intermediate-, and modern-epoch digitized all-sky photographic surveys. The astrometric and photometric quality of GSC–II (sub-arcsecond positional error, ~ 4 mas/yr absolute proper motion error, and $\sim 0.1 - 0.2$ mag photometric error) will make it a fundamental tool for telescope operations as well as for astrophysical investigations ranging from Galactic structure, dynamics and kinematics to extragalactic problems.

The Schmidt photographic plates on which GSC–II is based have a very non-linear intensity response; hence the availability of an all-sky network of photometric standards, spanning a luminosity range of about 10 magnitudes, is essential to the proper calibration of the photographic response function, i.e., for transforming measures of photographic density into a standard magnitude system. The construction of this new catalog of photometric reference objects, the Photometric Guide Star Catalog II (GSPC–II, Postman et al. 1996, Bucciarelli et al. 1998), while a larger project than either team could support alone, has been possible as a close and sustained collaboration of the OATo and ST ScI GSC–II personnel. The GSPC–II will provide calibration sequences in the V and R passbands (and in the B filter as resources permit, initially for about half of the southern sky) down to V=18–20, with photometric errors of ~ 0.05–0.1 mag. The goal is to provide each photographic plate with its own standard sequence lying within a circle of ~ 2.7° radius from the plate center, the purpose of this constraint being to avoid vignetted parts of the plate.

The data needed for the total of 1620 sequences which will constitute GSPC–II come from CCD observations made with different instruments and detectors, taken during a period of time spanning almost a decade. Such a huge amount of data calls for the use of a dedicated reduction pipeline, designed to meet the accuracy requirements and to handle data storage and retrieval for quality assessment efficiently. Accordingly, we have developed a new, highly automated, pipeline named ARLO (**A**utomated **R**eductions of **L**uminous **O**bjects), running under IRAF version 2.10.4[1]. Not only is ARLO adequate for all CCD data in our program, but we expect that it is sufficiently general for other CCD programs requiring a high level of automation.

This paper reviews the observing plan and the critical issues for precision photometry, then describes the pipeline in some detail, and finally demonstrates the kinds of error analysis and quality control supported in the ARLO environment. Then in the next paper of this series (Bucciarelli, Postman, et al. 1998) we shall present a detailed discussion of the photometric physics (transformations, extinctions, color relations, etc.), together with a first discussion of actual GSPC–II data and its quality.

2. Overview of CCD Observations

Generally, 1-meter class telescopes are used in this project, with only a few percent of the data coming from larger apertures. The detectors are different CCD cameras with sizes from ~ 3.5′ to ~ 23′ square, and with pixel sizes ranging from ~ 0.4″ to ~ 0.75″. The largest fraction of the northern data come from the 0.9 meter at KPNO, while in the south the observing was generally split between the the 0.9 m at CTIO and the 0.9 m Dutch at ESO La Silla, with $\delta \geq -30°$ and $\delta \leq -30°$, respectively.

A typical night of photometric data consists of a series of CCD exposures taken on program (GSPC–II) fields in the *Johnson-Kron-Cousins* B, V, and R filters, plus a fair number (> 20) of standard stars (Landolt 1992), imaged throughout the night in the same filters, and spanning a good range of air masses and colors. Each field is generally centered on the faintest star of the corresponding bright sequence for that field, as defined in the Guide Star Photometric Catalog I (GSPC–I, Lasker, Sturch et al. 1988). The typical exposure time to reach the desired magnitude limit with a good signal-to-noise ratio (~ 30) is 5-10 minutes for the V and R filters, and ~ 10 − 15 minutes for the B band. Occasionally, short (few minutes) exposures are taken for a program field; this usually happens when the faintest GSPC–I star present in the frame is saturated in the long exposure. Calibration frames, such as flat fields, bias and dark fields are also taken every night. Additionally, to monitor data quality, a few selected program fields are re-observed at every run, together with some open cluster fields for which literature photometry is available.

---

[1] ARLO is also operating under IRAF 2.11 at ST ScI

3. Critical Issues for Precision Photometry

A cornerstone of the ARLO goal to produce high quality pipeline (i.e., non-interactive) reductions is the requirement that two independent methods, *aperture photometry* and *PSF fitting*, give the same results. Aperture photometry, the numerical counterpart of classical photomultiplier photometry with a fixed aperture, is conceptually simple and gives reliable results in uncrowded fields, for objects well above the sky. PSF fitting, while more complex and requiring a separate zero point for each frame, is more robust in the presence of clutter and more effective in minimizing the effects of sky noise. Then the preferred GSPC-II reference objects are taken as those for which the two methods give consistent magnitudes, with objects having discordant photometry not generally being used.

As the detailed procedures for each of these methods is dependent on a number of parameter settings, the rest of this section is devoted to the arguments, simulations, and experiments underlying the selection of these parameters

3.1. Flat Field Calibrations

In order to achieve the goal of 5% standard magnitude error, one should be able to calibrate the CCD response to uniform illumination at the 1% to 2% level. Therefore, tests were conducted to compare different flat-field averaging criteria. We took five "good" sky flats in normal conditions (e.g. $\sim$ 30% of the saturation level) and also five "poor" flat-fields with low counts and visible stellar images. Then, the good flat-field was made using the first five frames (different combining methods did not produce relevant differences, i.e. deviations were smaller than 1 percent); and the poor flat-field was calculated using the second set of five frames, this time alternating all the combining methods. Using the IRAF parameters *average*, criterion for combining different frames, and *ccreject*, criterion for pixel rejection, resulted in the least difference between the "good" and "bad" flats. This was therefore taken as our method of choice, in agreement with the prescription given by Massey (1992).

The method for combining bias frames is easier to choose, since the only problem with these exposures is cosmic rays events. The best bias correction is obtained by rejecting the highest value from the given set, then averaging the remaining pixels. The IRAF **ccdproc** routine, called from ARLO, does the overscan correction and trims the image. It uses a Legendre polynomial to fit the distribution of pixels values in the defined overscan region. This allows a bias correction, even in absence of bias exposures. Additionally, when such exposures are available, they are used to perform a bidimensional correction over the entire frame. Dark exposures are treated, from a statistical point of view, the same way as bias frames, using the rejection of the highest value (IRAF *minmax* option) and averaging the remaining values.

3.2. Saturation

Saturated pixels must be corrected before any calibration is applied to the frame. Indeed, a saturated pixel (with value 65536 for a 16-bit analog-to-digital converter), could be lowered when divided by a flat-field value greater than unity – not an unusual condition – and treated as a "normal" pixel in the following procedures. It is necessary to put the value of saturated pixels (i.e. all the pixels above the linearity threshold) to such a high value that no flat field division could lower these pixels to a non-saturated condition (see Figure 1). ARLO accordingly sets the value of the saturated pixels to twice the nominal saturation threshold. Aperture photometry is not performed on stars with at least one saturated pixel inside the aperture circle. On the contrary, PSF photometry can be performed even if more than one pixel is saturated, provided that a sufficiently number of valid pixels are available to the fitting procedure.[2]

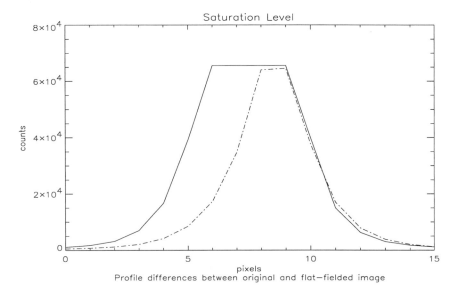

Figure 1. Profiles of a saturated bright star located in a region of changing sensitivity near the edge of the CCD. The solid line is from the raw image and the broken one from the flat-fielded image. The flat-fielded profile has changed substantially, with the shape near the edge (x axis origin) being greatly corrected due to the varying sensitivity of the CCD in that area, while the saturated flat top of the star has been lowered such that the data could erroneously be taken as unsaturated by subsequent software tasks inthe ARLO pipeline.

3.3. Cosmic Ray Detection

Cosmic Rays (CR) represent a serious problem for aperture photometry, especially when they lie near faint stars; for then total flux from the object is overestimated, with errors often greater than one magnitude. Figure 2 shows IRAF simulations for a set of stars contaminated by CRs. One notes the increasing photometric error when going to fainter stars. In the simplest case, these

---

[2] A possible improvement to the current pipeline could be to correct the saturated pixels using the profile obtained by the PSF photometry and then re-run the aperture photometry task.

bias the measures of individual stars; but if the CR contamination is on a star used to set the zero point for the PSF photometry, the results can be biased for an entire frame. In ARLO, CR are corrected using the IRAF task **cosmicrays**. To identify a CR, **cosmicrays** compares the flux ratio between the candidate CR and the surrounding pixels (in a 7 x 7 pixel box), tagging the object as a CR if that flux ratio is above a given threshold. The plots in Figure 3 show that the loci of stars and CR are well defined, such that a flux ratio of 5.0 is a good criterion to separate the two regions. CR are then removed using the mean flux of the surrounding pixels. Correcting the CR instead of just marking them as "bad pixels" allows to obtain aperture magnitudes that will otherwise be lost. In fact, IRAF does not compute an aperture magnitude when "bad pixels" are inside the photometric aperture radius.

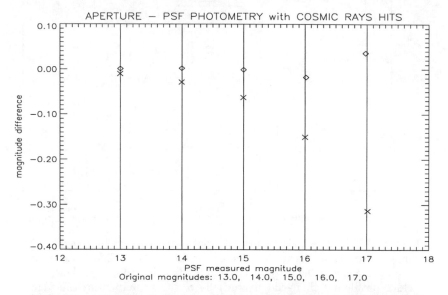

Figure 2. Departures of measured magnitudes for a simulated star hit by a Cosmic Ray. Diamonds represent the difference between the simulated magnitude and IRAF PSF magnitude, while crosses are with respect to the IRAF aperture photometry.

3.4. Choice of Aperture Radius

A delicate point with the aperture method is the choice of the radius. It is well known that the use of a large area, i.e., collecting practically 100% of the light, has the undesirable effect of increasing the sky noise with the square root of the aperture. The photometric literature suggests that an optimal strategy would involve the choice of the radius which gives the best signal-to-noise ratio, together with an aperture correction based on growth-curve methods (Stetson 1990) However, the application of such methods in an automated fashion can be difficult; and we have preferred to adopt a radius equal to 2.3 times the mean gaussian *Full Width at Half Maximum* (*FWHM*) of over the whole frame.

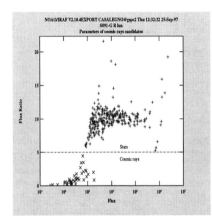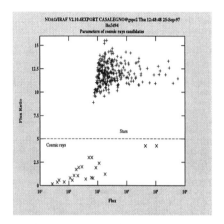

Figure 3. Flux vs flux ratio for representative CCDs, filters, and exposure times. The first panel refers to an 8 min exposure in the R, taken with CCD # 33 (27x27 μ pixels size) at the ESO Dutch Telescope; and the second panel is from a 40 sec exposure in the I filter, taken with the CCD (22x22 μ pixels size) available at the REOSC Astrometric Reflector at OATo. Crosses represent objects designated as CR, while the plus symbols are stars. Note that the stars and cosmic rays are clearly separated. Points near the border (at ratio 5) have been inspected visually to verify that no erroneous assignments have been done.

Additionally, for the case of very good seeing we limit the radius to 4.5" (see Massey et al., 1989) in order to encircle the seeing-independent part of the light distributed in the profile wings.

In practice, ARLO allows some flexibility in computing aperture photometry using different radius, and our standard configuration uses an aperture radii equal to $MAX(4.5/ccdscale, 2.3 \cdot FWHM)$. The IRAF parameters *daophot.fitskypars.annulus* and *sannulus*, representing the inner radius of the ring in which the sky background is computed, are both set to $MAX(2.61 \cdot 4.5/ccdscale, 6 \cdot FWHM)$, 2.61 being the proportionality factor between 6 (the coefficient of the $FWHM$ used to compute the sky ring) and 2.3. The reason for such a large value is that it is important to have a sufficient number of pixels in the statistics, avoiding interference from field stars. With such a radius (usually 10 pixels), tests have shown that up to six nearby stars (objects, in general) do not to produce significant variation of the sky value.

3.5. PSF Star Selection

The automatic selection of stars to be used for PSF fitting is a critical ARLO configuration issue. This selection is done using the IRAF task **pstselect**. In its normal mode, this task checks only for nearby stars brighter than the PSF candidate star; however, our tests show that this can result in a distortion of the PSF Look-Up Table (LUT) due to the presence of fainter companions, thereby biasing the final PSF magnitudes. Accordingly, we correct the rejection criterion so as to exclude any PSF star with a nearby companion (previously detected by the IRAF task **daofind**) up to 4 magnitudes fainter.

Galaxies represent a problem if they are erroneously used to estimate the stellar profile, especially in the analytical core. The IRAF task **psf** needs one star to calculate the parameters of the fitting function, plus a set of other stars to obtain an empirical correction, stored in the LUT. The ARLO default number of stars is 5 to compute the LUT, plus one for the analytical component. If galaxies, incorrectly classified as stars, are included in this set, they can deeply affect the analytical and/or empirical part of the fitting function, as illustrated in the first panel of Figure 4, where a galaxy with a small ellipticity has been used to compute the analytical parameters of the PSF, while two galaxies plus two stars defined the empirical correction. It can be seen that the wings of the function are biased due to the different shapes of galaxies and stars. The second panel of Figure 4 gives the difference between "true" magnitudes (i.e., simulated) and the computed *PSF* photometry for a set of five stars (the objects used to compute the PSF are not included in this plot). The stellar PSF magnitude is heavily overestimated, especially at faint magnitudes.

While the probability of a galaxy being picked by the PSF star selection task is very low, as the candidate objects are ranked in order of decreasing brightness, for particularly poor fields this situation can be a concern. A useful test in this case is to compare aperture and fitting photometry of the objects used to define the PSF profile. The degree of concordance of the two photometries is utilized by the PSF selection task as criterion for rejecting dubious objects, as explained in the following section.

3.6. Zero Point Determination

Theoretically, the PSF volume of a point-like source is proportional, within a single frame, to the total flux coming from it; nevertheless, this proportionality is not preserved from one frame to another, essentially because the analytical model never exactly reproduces the *real* stellar profile. It is therefore necessary to link the PSF photometry coming from different frames to a common magnitude system by setting a *zero point* in the equation

$$m = zp - 2.5 \log_{10} vol$$

where *vol* is the volume enclosed by the analytical profile fitting, m is the magnitude, and zp is the zero point. By default, IRAF task **psf** computes the zp by attaching the volume *vol* of the first star selected (to compute the analytical PSF) to the corresponding magnitude calculated by the aperture photometry task. All the other volumes in that frame can be converted into magnitudes using this zp. Problems arise when the star used for the comparison is not a

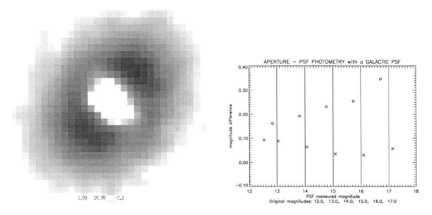

Figure 4. The first panel shows a Look-Up Table as obtained from a "bad" set of objects (stars mixed with galaxies), and the second shows simulated magnitude errors on stars reduced using a defective *PSF*.

"good" star, e.g., it is a double, or another bright object lies inside the PSF fitting radius, or it is a misclassified galaxy. In ARLO, the zp found with the first star selected by **pstselect** is checked comparing the PSF photometry of the other stars in the list (usually 5 stars) with their aperture magnitudes. If the median of the differences is higher than 0.015, the first star is discarded (on the supposition that it is a "troubled star"), and the next in the list is used to set the zp. The procedure is iterated up to six times or until the shift is smaller than the chosen empirical threshold. The final shifts are written in a table and used for statistical purposes. Experience based on real data have showed that this approach truly improves the performance of the automatic PSF star selection task.

4. CCD Reductions

The ARLO pipeline for the reduction of GSPC-II photometry sequences consists of a series of tasks written as standard IRAF *scripts*. As such, while the dominant experience is with IRAF 2.10.4 under linux at OATo, ARLO should run under any standard IRAF installation, and therefore is offered to the community as general purpose photometric reduction package, which can be obtained (together with a detailed user-manual, Casalegno 1998) by following the GSPC-II links from **www.to.astro.it**.

A driving requirement in the ARLO development was the on-line availability of all data relevant to the GSPC-II error analysis. This is accomplished by storing all descriptive information output from various IRAF tasks as STSDAS tables, which can later be used both for off-line analysis and as inputs to new tasks. A description of the main output files produced by ARLO, as well as a flow diagram of the pipeline, is given in Figure 5.

The main ARLO tasks are described in the remainder of this section. The user not fully familiar with the principles of CCD photometry may also wish to consult Howell (1992).

4.1. Flat-Bias-Dark (FBD) Task

This is the first task run by the pipeline. It translates FITS format images into IRAF format, cleans the images, (bias, dark and flat correction), updates image headers with the necessary keywords, and creates the lists of image filenames which are grouped according to type and filter used. The origin of the FITS file (site/telescope/detector) is identified by comparing the header with those stored in a user database; this allows ARLO to apply the correct calibration to the frames.

4.2. Parameter Auto Detection (PAD) Task

This task automatically determines the image-processing parameters. While some quantities are (and *have* to be) kept constant for the entire set of frames, others may change from frame to frame. An interactive approach (i.e. with IRAF task **daoedit**) is time-consuming and boring, while **PAD** is fast and completely automated. It calculates the mean background and corresponding standard deviation for every frame. This is done iteratively, by applying a 3-sigma rejection criteria until no more pixels are rejected (or up to 50 iterations). Cosmic rays are removed using the IRAF task **cosmicrays** (see section 3.3).

To detect the stellar parameters, **daofind** is run with some specified input values; in particular, the *sharpness* limits are set to 0.2 to 0.6. The sharpness is defined as $(peak- < pixels >)/height$, where the numerator is the difference between the central pixel value and the neighboring pixel average, and the denominator is the height of the best gaussian fitting at that point. Another parameter is the *roundness* value, computed fitting a one dimensional gaussian function to the x and y marginal pixel distribution, and computing the value $2(hx-hy)/(hx+hy)$, where hx and hy are the peaks of the fitted gaussians. The default ARLO limits for this parameter are -0.5 and 0.5. Then, the *median* of the individual gaussian *FWHM* is calculated by the IRAF task **imexamine** in an automated way. Several tests have shown that the use of the above thresholds gives a robust detection of the frame parameters.

4.3. Aperture and PSF Photometry (APP) Task

This task performs *both* Aperture and Point Spread Function photometry. It steps through several IRAF packages (**daofind, phot, pstselect, psf**, and **allstar**), reading as required the parameters calculated by **PAD**. It does not require user assistance, but it is the slowest task in the ARLO package due to the complexity of the computing methods. The first operation performed by **APP** is finding the objects; **APP** uses a threshold set by default to 4 (user parameter) to detect all possible star candidates in the frame. In **daofind**, the *sharpness* values are relaxed to the limits -0.2 and 1, while the *roundness* value is considered good from -1 to 1. The task outputs a list of objects to a file, subsequently used by **phot** and **allstar**, which calculates aperture and PSF instrumental magnitudes respectively. The aperture magnitude is computed as

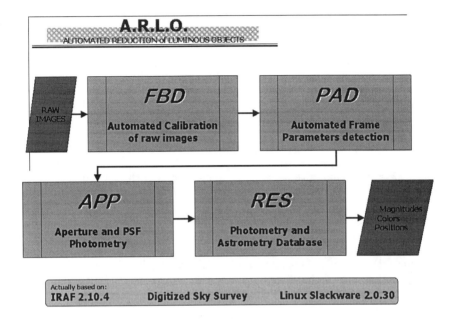

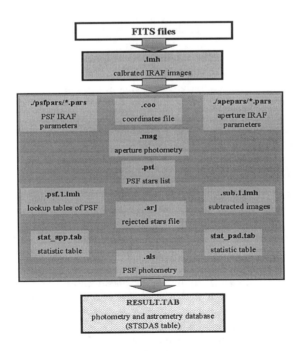

Figure 5. Flowchart and files from ARLO pipeline

$$magnitude = zero - 2.5\log_{10}(sum - area \cdot sky) + 2.5\log_{10}(time) \quad , \qquad (1)$$

where *time* is the exposure time, *area* and *sky* are respectively the area of the sky annulus and the estimated sky value (in counts), and *sum* is the sum of the pixel values within a radius equals to the aperture chosen (depending on the method chosen), with fractional pixels and no weighting techniques. The gain of the CCD is used to compute the magnitude error, which is calculated by

$$merr = 1.0857 \frac{\sqrt{\frac{sum-area \cdot sky}{gain} + area \cdot stdev^2 + \frac{area^2 \cdot stdev^2}{nsky}}}{sum - area \cdot sky} \quad , \qquad (2)$$

where *merr* is the magnitude error, *sum* the counts within the aperture radius, *stdev* and *nsky* the standard deviation of the *nsky* pixels in the sky annulus. The first term inside the square root is an estimation of the photon noise error due to the source, the second represents variations of the mean sky, while the third accounts for the error of the mean sky subtracted from the source. For stars partially out of the image, with bad pixels or when the sky can not be computed, the magnitude is given an INDEF value. The sky value is computed taking the median of the distribution of the counts inside an annulus around each star, of inner radius equal to $6FWHM$, and with width usually equal to 10 pixels (see section 3.4).

The best representation of the PSF for a generic CCD frame is chosen to be a combination of an analytical core described by one of the built-in IRAF function (the *Penny*1 function is the generally preferred one), plus a look-up table which allows for deviation of the outer PSF profile from the analytical representation of choice. The PSF algorithm computes the photometric error in two ways: (1) from the readout noise and gain of the detector, the standard error of the flux in each pixel is calculated, and this is propagated through the least squares solution to predict the standard magnitude error; (2) after the fit has been performed, the *rms* value of the observed pixel residuals is computed. An average of these two estimates is returned as the standard error of the magnitude (see Stetson 1987 for details).

4.4. RESult (RES) Task

The final pipeline task, which is preceded by standardization of the instrumental magnitudes (performed running the IRAF tasks **fitparam** and **invertfit**) is concerned with the creation of STSDAS tables for the storage of all useful information produced by the previous IRAF tasks. In particular, for each frame a table containing *object* data, such as image features, pixel coordinates, sky background, etc., is generated. For each observing night the pipeline produces also a table with *frame* data, which includes, e.g., seeing quality, average sky background, number of objects detected in each frame. These data are kept on-line and used for statistical purposes and quality assurance of the catalog. Astrometry of all the CCD objects for which photometric data have been computed is also carried out at this stage, by means of an ARLO sub-task, as described in the following section. The last step consists in the creation of a main ASCII table containing the final standard photometry and astrometry, plus instrumental photometric data, output of the IRAF tasks **phot** and **allstar**. These data are used to populate the so-called *intermediate* field catalogs, which, once appropriately screened, will ultimately constitute the GSPC-II catalog.

4.5. Computing of Coordinate System (CCS) Task

This pipeline task is based on the IRAF routine **ccxymatch**, which works on sky- and frame-based coordinates, using the *triangles* matching algorithm (Groth 1986). The astrometric solution is based on the ST ScI Digitized Sky Survey (DSS), from which cutouts centered on each CCD frame have been previously extracted and from which lists of DSS stellar objects with positions and fluxes have been generated. The task performs the following steps: first, a match between the DSS (α, δ) coordinates and the CCD (x, y) coordinates is computed, then a plate solution is carried out by using a linear transformation between matched objects, and finally the plate solution is inversely applied to all the remaining objects in the frame. An error equal to the error of the *rms* of the fit is also attached to each coordinate. If for some reason the automatic match fails – this can happen when there are too few stars in the DSS cut-out, or when the field is too crowded, or for some odd stellar distribution – the procedure displays both the CCD image and the DSS cut-out on the workstation, and the user is asked to manually identify a few stars on both images, with which a plate solution can be computed. Our experience is that the *triangles* algorithm, which represents the core of the astrometric pipeline task, is very sensitive to the distribution of the stellar field, its population, and in particular to the ratio between the number of objects in the two lists to be matched. Therefore, when necessary, the original lists of objects are cut down in size in order to increase the success rate of the match. Moreover, for very poor/faint stellar fields (less than 5-6 stars) the automatic procedure can give a false match. To avoid this situation, the pipeline performs a quality check of the transformation parameters by comparing them with the nominal CCD scale and rotation values, and when a false match is detected, the procedure switches to manual. After these optimizations, the automatic part of the tasks takes care of about 80% of the cases.

5. Discussion of Photometric Errors

A very important feature of ARLO is that both *aperture* and *PSF* photometry are carried out. This permits us obtain rigorous monitoring of the data quality for every frame by making direct comparisons of the two photometric reductions. It is known that where a good PSF function can be found for a particular exposure, the resulting instrumental magnitudes have smaller errors at faint magnitudes than their counterparts obtained with the classical aperture method (c.f., Figure 8). This is because the sky noise increases as the square of the area over which the stellar flux is collected in the case of aperture photometry, while the PSF profile fitting has the effect of minimizing the sky-area that enters into the reductions. On the other hand, an inferior set of PSF stars can lead to systematic errors in the photometry, which, again, can be controlled by comparison with the aperture method; and in such cases we need to use the aperture results, even though their errors are larger.

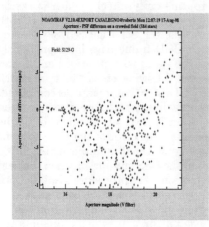
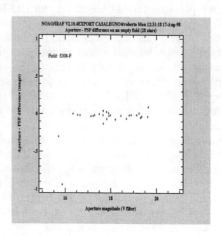

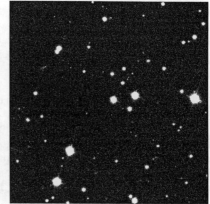

Figure 6. Left panels (a):*Aperture* versus *PSF* photometry and CCD image (600 sec exposure) of field S129-G; Right panels (b): same for field S308-F (900 sec exposure)

Moving from simulations to real data, Figure 6 presents differences in the two photometries for two very different stellar fields: one at intermediate Galactic latitude $b \sim 20°$ and the other near the Galactic plane. It can be seen that for a "well behaved" stellar distribution, as in the case of GSPC-II field S308-F (Fig. 6b, intermediate latitude), both methods give good results, with agreement at the level of a few percent; and the discrepancies mostly indicate wrong aperture magnitudes due to nearby-objects interference. For the other field, S129-G (Fig. 6a, low latitude), which is densely populated, multiple effects come into play and the interpretation of the residuals is more complicated. The majority of the negative residuals are caused by spurious objects falling inside the aperture radius. A bias can also be noted at the very faint magnitude end (see the tilted edge of the graph), possibly due to a systematic failure of the fitting procedure under certain conditions; this is typical of the situations requiring further investigation before adopting final photometric values.

We emphasize that the GSPC-II goal is to include only CCD objects of the highest quality in the final catalog. Typically, this means (a) consistency of aperture and PSF photometry (b) and sufficiently clean images that they can be reliably used as calibrators in the GSC–II photographic photometry. And naturally, there will be manual screening when these criteria are not directly achievable.

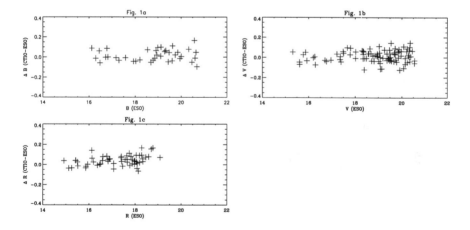

Figure 7. Differences in B, V, and R PSF magnitudes for GSPC–II fields S313 and S442 imaged with the ESO Dutch telescope + CCD Tek #33 and the CTIO 0.9 m telescope + CCD Tek1024. Only stellar magnitudes with formal errors smaller than 0.05 have been selected. The average *rms* values (strictly, these would be a function of magnitude) for the three filters are equal to 0.06 *mag*, which is quite satisfactory.

The well-structured and self-documenting characteristics of the ARLO data are also conducive to a broad variety of other tests. One example (Figure 7) is the use of common control fields to evaluate the consistency of the data from night to night and between observatories. The next paper of this series (Bucciarelli, Postman, et al., 1998) will use these capabilities, together with a major subset

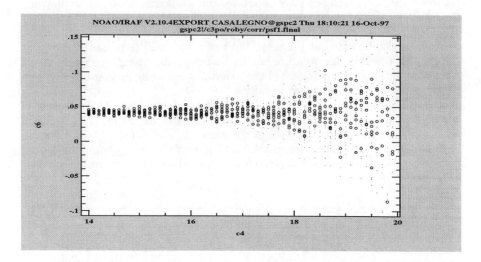

Figure 8. Representative precision of aperture and PSF photometry, as shown via a simulation of 30 artificial frames with 60 stars each. The magnitude range is from 14 to 20, with bins of 0.1 magnitude. Dots and circles represent the deviation from the "true" magnitude of the aperture and PSF photometry, respectively. The plots show how the performance of aperture photometry deteriorates at faint magnitudes, due to sky background contamination.

of the reduced GSPC–II data to present a realistic view of the data quality for the actual GSPC–II catalog.

Acknowledgments. We wish to thank the OATo and ST ScI teams, and in particular C. Sturch, for making their expertise available to us during this work. We also acknowledge ESO, CTIO, and KPNO for continuing support with telescope time.

The work at OATo has been partially funded by ESA/ECF, contract number 52899/ECF/98/6327/HWE, and by the Italian Council for Research in Astronomy. The Space Telescope Science Institute is operated by the Association of Universities for Research in Astronomy, Inc., under contract to NASA.

References

Bucciarelli, B., Postman, M., Lasker, B.M., Casalegno, R., Garcia, J., Sturch, C., et al. 1998, in preparation.

Casalegno, R. 1998, *The ARLO User Manual*, OATo Internal Report No.44.

Groth, E.J. 1986, AJ, 91, 1244.

Howell, S. ed. 1992, *Astronomical CCD Observations and Reduction Techniques*, ASP Conference Series vol. 23, 1992.

Landolt, A.U., 1992, AJ, 104, 340.

Lasker, B.M., Sturch C.R. et al. 1988, ApJS, 68, 1.

Lasker, B.M., McLean, B.J., Jenkner, H., Lattanzi, M.G., and Spagna, A. 1995, in *Future Possibilities for Astrometry in Space*, ESA SP-379, p. 137.

Lindsey, E.D. 1994, *A Reference Guide to IRAF/DAOPHOT Package*, N.O.A.O. IRAF Programming Group.

Massey, P. 1992, *A User's Guide to CCD Reduction with IRAF*, http://iraf.noao.edu/iraf/web/docs/docmain.html

Massey, P., Garmany C.D., Silkey M., Degioia-Eastwood K. 1989, AJ, 97, 107.

Postman, M., Bucciarelli, B., Sturch, C.R., Borgman, T., Casalegno, R., Doggett, J., Costa, E., 1998, in *New Horizons from Multi-Wavelength Sky surveys*, IAU Symp. 179, p. 379.

Stetson, P.B. 1987, PASP 99, 223.

Stetson, P.B. 1990, PASP 102, 932.

Photometric Search for Extra-Solar Planets

Steve B. Howell, Mark E. Everett

Wyoming Infrared Observatory & Department of Physics and Astronomy, University of Wyoming, Laramie, WY 82071
showell@mothra.uwyo.edu; everett@kaya.uwyo.edu

Gilbert Esquerdo, D. R. Davis, S. Weidenschilling

Planetary Science Institute, 620 N. 6th Street, Tucson, AZ 85705
esquerdo@psi.edu; drd@psi.edu; sjw@psi.edu

Trish Van Lew

Department of Physics and Astronomy, University of Wyoming, Laramie, WY 82071
jibben@tana.uwyo.edu

Abstract. Searching for planets orbiting other suns has become an area of intense interest in recent years. Most search programs use high precision radial velocity methods and use long term, detailed monitoring of a few bright F,G stars. Photometric methods have been briefly considered over the last few years, but are generally thought to be unable to reach the needed precision levels. We present here an initial report on the development of a low-cost, photometric search program telescope. Our inexpensive system is capable of producing ultra-high photometric measurements, currently yielding standard deviations as low as ±0.004 magnitudes or 0.4% photometric precision. However, a number of limiting factors resulting from both the equipment and from our current lack of complete understanding of ultra-high precision photometry have been identified. We show that with a system such as described herein, detections of planetary transits by Jupiter-sized bodies in orbit around F-M dwarfs could be accomplished.

1. Introduction

One of the most important astronomical pursuits of this decade has been the search for planets encircling other stars. The discovery of such objects has had a profound influence on both scientific and non-scientific disciplines. Numerous religious and philosophical issues have once again been raised to the forefront of human thought. While these topics are exciting and of much interest, we refrain from further discussion of them in this treatise.

Planets in orbit around other suns is certainly not a new idea. Epicurus mentioned this possibility as early as 300 B.C. but the idea lost favor with the coming of the "earth-centered" view held for 1500 or so years. The idea of planets orbiting stars was laid on firm ground when Copernicus placed the sun in its rightful place at the center of our solar system. Indeed, this Helio-centric view held up and was not seriously questioned again. But what about actual strides to suggest that extra-solar planets might exist or possible methods of searching for them?

In the landmark series *Lardner's Handbooks of Natural Philosophy and Astronomy* (1858), Dionysius Lardner provided the scientific minded citizen of the time with a complete set of works containing the essentials in all manner of physical science. Volume three of the 3 part series, the Third Course: Meteorology and Astronomy, contained almost 700 pages of text dealing with the stars, planets, and many other areas of astronomical knowledge. In Chapter XXVII we find the discussion of periodic variable stars. Five suggestions were made as to the cause of such a phenomena. Starspots, conflagrations due to comet impacts, precessing disks, and dense atmospheric clouds are all discussed in short, vague detail, but it is the fifth suggestion which gets the author's vote. *"Periodical obscuration or total disappearance of the star, may arise from transits of the star by its attendant planets."* The paragraph goes on to discuss the fact that none of our sun's planets could render it a periodic variable as seen from afar due to their small size. Thus, the argument led to extra-solar planets being of either very large size, comparable to their sun in mass, or being of very small density. It was even further suggested that in some cases, the central body is a planet attended by a lesser sun.

By 1900 however, most college astronomy textbooks (ie., Elements of Descriptive Astronomy by H. Howe, 1909) had dropped the idea of planets as a possible cause of variability, instead giving more credence to areas for which observational and theoretical progress had been made; starspots, eclipsing binary stars, and tidal disturbances caused by massive stellar companions.

Today, there are nine probable planets encircling other stars, most having minimum masses equal to that of Jupiter or larger, with two likely to be about 0.5 M_J. Marcy et al. (1997), Cochran et al. (1997), Marcy & Butler (1997, 1998) describe in detail the current situation in extra-solar planet research. High-precision spectroscopic studies of minute changes in a star's radial velocity have been used to detect all of the currently suspected extra-solar planetary systems. Detailed radial velocity work requires months to years of study for each star and can only be accomplished currently on a few of the world's large telescopes which are equipped with spectrographs of sufficient stability and precision. As such, radial velocity study is a painstaking method requiring great care, but the rewards make it worthwhile, yielding the planets' orbital periods and a mass estimate (albeit uncertain by the sine of the orbital inclination). Radial velocity search methods are currently confined to detecting massive planets, such as Jupiters, in orbit around bright F,G stars. With the availability of 8-10 meter class telescopes, searches for smaller mass (radii) planets and planetary systems in orbit around later spectral types, are clearly avenues awaiting exploration.

The theory of photometric transit searches has been postulated and discussed by a number of authors (e.g., Rosenblatt 1971; Borucki & Summers 1984; Schneider & Chevreton 1990; Howell & Merline 1995; Giampapa et al. 1995; Howell et al., 1996) These authors present numerous details of the theory and methods of extra-solar planet search techniques. The photometric transit search method is not only sensitive to planetary systems such as those observed by current spectroscopic methods (Jupiter-sized planets in orbit around F-G dwarfs), but is also the only current method which is capable of detecting terrestrial-sized planets. Transits by planets orbiting F-M dwarfs can be detected, with K-M dwarfs allowing earth-sized bodies to be possibly detected even with modest sized ground-based telescopes. Photometric transit searches are the prime or secondary scientific objective for the approved COROT and the proposed KEPLER (formerly FRESIP) satellite missions and a number of other smaller endeavors.

In this paper we describe our investigation into the feasibility of using wide-field photometric monitoring campaigns to search for transits by extra-solar planets. We have established a pilot project with goals of understanding methodologies of ultra-high precision photometric techniques, investigation of long-term wide-field differential photometry, detection of transits by extra-solar planets, and study of previously unrecognized low-level photometric phenomena such as new types of variable stars. In §2, we begin with a short review of the subject of transits, their specifics of interest here, and a brief discussion of the detection probabilities. In §3, we discuss our pilot project, observing setup, and observing plan, §4 presents initial results, and we end with some conclusions about the possible results and implications of this type of work for extra-solar planets and its complementary nature to other search methods.

2. Photometric Transits

A transit by an extra-solar planet (ESP) will occur, for a fortuitously aligned orbital plane, as that planet passes in front of its parent star. During the transit, a small decrease in light from the parent star will occur with an associated although small color change due to limb darkening; see Borucki & Summers (1984) and Giampapa et al. (1995). Measurement of the time of light decrease (ingress), total transit duration, and the time of light increase (egress), all provide information about the transiting object. These times can be used to constrain the orbital period of the planet while the amount of decrease can tell us the size of the planet. By using reasonable densities for a given planetary size, a mass can be inferred. The decrease in light is in direct proportion to the ratio of the projected areas of the planet to the star. Formally, this is

$$\frac{\Delta I}{I} = \left(\frac{R_{Planet}}{R_{Star}}\right)^2$$

For example, if we consider the Sun and Jupiter the decrease in signal is 1% (0.01 mag) and with Jupiter's orbital parameters the duration of a central transit would be ~ 30 hr. For Neptune, the decrease is 0.1% (0.001 mag) for 71 hr; and for the Earth, we have 0.01% (0.0001 mag) for 13 hr. Transit times (for a central crossing and a circular orbit) can be determined from the formulation

$$t = 13 R_* \sqrt{\frac{a}{M_*}}$$

where R_* and M_* are the stellar radius and mass in solar units respectively, a is the radius of the planet's orbit in AU, and t is the transit time in hours. For relatively short period orbits, e.g., a few days to weeks, the planetary orbital period may be obtained by observing the star for many nights, yielding multiple transits.

Table 1 provides example transit properties for star-planet combinations with planetary orbital radii that scale proportionally to the radii of the parent stars. Two cases are considered; one which matches the Sun-Jupiter system and one for 51 Peg-like systems. We list orbital periods, transit times, and the decrease in light for each combination. We note here that the required photometric precisions to actually observe a transit by an extra-solar planet are remarkable, but easy to obtain with today's modern CCD detectors and differential ensemble photometric techniques (e.g., Howell & Merline, 1995; Gilliland et al. 1993).

Table 1. Photometric Transit Properties for a Jupiter-Sized Planet

Sp. Type	Scaled Solar-like System		Scaled 51Peg-like System		$\Delta I/I$
	Orbital Period (years)	Transit Time (hours)	Orbital Period (days)	Transit Time (hours)	(mag)
G2V	12	30	4.2	2.9	0.010
K2V	10	25	3.6	2.5	0.015
M2V	6.7	17	2.4	1.7	0.040

For example, the best photometric precision one could hope for would be to reach the scintillation limit of the earth's atmosphere. At one airmass, this has a typical value of 0.0004 mags or 0.04% (Young 1974). Using a CCD (or in fact almost any detector) the best obtainable precision will occur if all of the collected photons are from the star and no other noise source contributes. In this case, the photometric error is approximately given by the inverse of the square root of the total signal, in accord with pure Poisson statistics. There is actually a small correction factor between the error in electrons and the error in magnitude, and the actual 1σ error in magnitudes is given by (Howell 1993)

$$\sigma_* = \frac{1.0857 \cdot \sqrt{N_* + p}}{N_*}$$

which reduces to

$$\sigma_* = \frac{1.0857}{\sqrt{N_*}}$$

in the case of detection of only source photons. Here N_* is the total number of photons detected from the star and p is the total number of photons or electrons contributed by all sources of noise.

Thus, we see that the more source photons you can collect, the better your final photometric precision can be. However, other noise sources such as sky background, CCD read noise, and systematic effects can greatly reduce the final precision below the optimum level. The use of differential techniques essentially eliminates all such noise and systematic terms and allows one to obtain results near to or even at the theoretical level (see Howell et al. 1988).

To detect and measure extra-solar planet transits, we require high photometric *precision* rather than *accuracy*. In other words, knowledge of the observables $\Delta I/I$, transit duration, and planetary orbital period depends only on the internal uncertainties in the light curves (e.g., the noise) and not the absolute calibration to a standard photometric system. In fact, we are presently operating our pilot system telescope without a filter (white light), allowing the bandpass to be mainly determined by the CCD quantum efficiency curve. This yields an approximate V+R-band magnitude for our example data which we have roughly calibrated to V magnitude using observations of photometric standard stars. Our quoted V magnitude thus gives a rough indication of the brightnesses we are observing. We show in §4 that the ensemble differential photometry techniques we use, do indeed produce high precision light curves. Accurate magnitudes, colors, and classification spectra can be obtained either before or after detecting a transit candidate, or possibly during a later, predicted transit.

To detect a transit we require that as the stellar flux decreases (and its magnitude increases), the change in magnitude, Δm, can be significantly detected given the uncertainty in our individual differential magnitude estimates σ_*. If only one magnitude measurement were made during a transit, a relatively large deviation from the mean magnitude would be required (e.g., perhaps $\Delta m \geq 5\sigma_*$). However, we expect transits to last for hours (see Table 1) and thus would have multiple magnitude measurements for each transit. With many nights' data for each star in the field, we would also know mean stellar magnitudes very precisely and have a long temporal record of the characteristic variability of each star. If N magnitude measurements are made during a transit, we could detect transits with smaller values of Δm (e.g., $\Delta m \geq 5\sigma_*/\sqrt{N}$). Since we expect that short-period orbit planets would dominate our detections (see discussion below), we are likely to observe planets that make multiple transits during our survey. This would further increase the significance of our detections and permit us to determine the orbital period as well.

Having discussed transits, their photometric detection, and their use for determining orbital periods and sizes (masses) of ESP's, we now turn to a brief discussion of the probability of actually observing a transit. For the Sun-Jupiter system viewed from a random direction, the probability that the orbital inclination is near enough to the line of sight ($< 0°.1$) to allow observable transits is $\sim 10^{-3}$. Thus, one would have to monitor at least 1,000 such stars for an interval of one planetary orbital period in order to see even one transit. Equivalently, the fraction of stars with transits occurring at any instant is the fractional solid angle subtended by the star as seen from the planet. This is $\sim 10^{-7}$ for the Sun as seen from Jupiter. However, the orbital radius of systems such as 51 Peg ($\sim$one-third of all systems in which planets have been found thus far) is about 100 times smaller, meaning that the probability of seeing a transit is $\sim 10^4$ times larger. Such (inner) planets will dominate detections by all methods, even if they are present in only a small fraction of ESP systems. Additionally, if a typical star has multiple planets, the transit statistics will be dominated by the innermost planet but will increase overall for that system.

Since planetary transits are easier to discover as the stellar radius decreases and we believe that solar-like stars (F-M) are better search targets than earlier spectral types, we will pick search fields in an intelligent manner. Fields which are too sparse or too crowded are avoided due to overlapping images and small total sample respectively. Early-type stars (O-A) are also avoided due to age, spectral type, and CCD saturation considerations. Thus, an areal photometric search near selected mid-galactic latitudes ($20° - 40°$) generally provides a good compromise. At these latitudes and given our selection criteria, the search fields used provide a magnitude-limited sample which contains mostly late-type dwarfs with equal or smaller radii than the Sun (Bahcall & Soneira 1980). Figure 1 illustrates the necessary photometric precisions to detect Jupiter-, Neptune-, and Earth-sized planets in orbit around F-M dwarfs. We note here that Jupiter provides the maximum effect that can reasonably be expected for a planet. The equation of state of hydrogen is such that more massive planets, if they exist, will not be significantly larger than Jupiter, only denser. Any larger object must have an internal heat source, e.g., a brown dwarf or close stellar companion.

3. Photometric Search System

Our prototype ESP search system is based on "off-the-shelf" components which we melded into a single system. Primary drivers of this choice of system were that we wanted to test if a low cost system was capable of obtaining the desired precision for this type of project and the entire project was to be run almost entirely by undergraduate students. We also did not want to spend time and money, at this stage, to develop custom software components for data acquisition, reduction, or analysis. In addition, duplication of such a search system can be easily accomplished.

3.1. Telescope + CCD

Our choices for a telescope and CCD were points of much debate in the early phases of the project. It was necessary to select a system that would be adequately responsive such that integration times of a reasonable duration (to sample transit ingress and egress) could be used (2-4 minutes), yet also provide a large enough field-of-view (FOV) in order to obtain a sufficient sample of stars. Additionally, choices of low-cost telescopes and CCDs are very limited, especially if a detector of good scientific value is desired. We also needed to consider the software available to run both the telescope and CCD, a choice driven by our desire to operate from a PC class machine. Our final choice of items for this prototype system are listed in Table 2 along with their approximate purchase prices.

Table 2. Pilot Telescope + CCD System Components

Component	Description	Price
Telescope	Meade LX-200, 8" f/6.3 Schmidt-Cassegrain	$2900
Accessories	Meade 8" wedge,	$160
CCD	Santa Barbara Instrument Group (SBIG) Model ST-8[a]	$6500
Computer	Compaq Deskpro 200 Pentium 166	
	2.5 GB hard drive, 32 MB Ram	$1580
Accessories	Windows NT, 14" monitor	
	Hayes 33.6k Modem, Writable CD, misc.	$1000
Optics	Celestron f/6.3 focal reducer-image corrector	$130
Software	Software Bisque's CCDsoft	$200
Misc.	Cables, CDs, telescope cover	$500
Total Cost		$12,970

[a]Kodak KAF1600 CCD with 1530×1020 9μ pixels.

The largest drawbacks to our system of "off-the-shelf" components are: (i) they are not designed to perform scientifically precise measurements nor do they have the needed capabilities for such (for example the CCD software has numerous canned routines for forming pretty pictures, but essentially no detailed routines for items such as plotting of pixel values within a point-spread function or even displaying the entire range of image data) and (ii) the associated manuals and manufacturer technical support are very poor in terms of any scientific issues (for example, the gain of the CCD camera is almost impossible to find out, we measured it ourself). Numerous other minor problems and issues were encountered but dealt with or worked around.

In order to obtain as large a field of view as possible, we investigated using 1 or 2 focal reducers with our system. Two focal reducers were chosen to expand the FOV to $1°.3 \times 0°.85$, giving 3 arcsec pixel^{-1} at an effective focal ratio of f/3.0. This arrangement provided reasonable star images (FWHM $\sim$ 3 pixels) in the central 1° circle of the FOV. Outside this circle, however, the field became highly vignetted and the stellar point spread functions (PSF; see King 1971, Diego 1985) appeared quite comatic. This is not an unusual occurrence for a wide field imager (Howell et al., 1996).

Current work with our system makes use of a single focal reducer or none at all. The specifications with one focal reducer are a FOV of $1°.0 \times 0°.65$ (giving 2.3 arcsec pixel^{-1} at an effective focal ratio of f/4.0. With this arrangement, the star images improve to a FWHM of $\sim$ 2 pixels and there is no noticeable vignetting in the field. Without a focal reducer the FOV is $0°.62 \times 0°.41$ (giving 1.45 arcsec pixel^{-1}) at f/6.3. We are currently taking images in white light. Investigation of use of a broad-band V filter produced no noticeable change in the PSF, although it did decrease the total system throughout.

For our SBIG ST-8 CCD we have measured a gain of 2.7 electrons/ADU, read noise of 11 electrons rms, and dark current of 0.7 electrons/second at a nominal operating temperature of $-15°$ C. The data is read out from the CCD as 16 bit integers, but unfortunately the commercial software can only display the FITS images using 15 bit (signed) integers. This greatly limits the usefulness of any "at the telescope" inspection or analysis of the CCD data. Confusion related to the actual count levels obtained within a specific star is also possible under such conditions. Once the data is transferred to the IRAF environment (see Section 3.3) it is stored as full 32-bit data, thus viewing of the full data range is available.

3.2. Data Acquisition

The LX-200 telescope is controlled through an on-board computer which contains a database of nearly 65,000 pre-programmed objects. For our purposes, we utilize the option of entering any desired RA and Dec. After initialization of the digital encoders at the beginning of the observing session, the telescope will slew to any entered object or position. This is quite useful for locating our search fields. (Note that typical search fields have few bright stars [V<9] as we wish to avoid saturation.) Commands to slew the LX-200 are entered through a small hand paddle attached to the telescope, eliminating the need for an external computer for telescope control (though external computer control is an option).

The CCD camera is operated via the *CCDSoft* program written by Software Bisque. The CCD is instructed to take time-series data throughout the night, automatically saving each frame to the hard drive of the control computer (a Pentium based PC). Initially, it was found that, as shipped, CCDSoft could not conduct these tasks. Careful discussion with the software engineers at Software Bisque led to minor modification of the configuration files necessary to run the system. As a result, the system can now be operated in time-series mode, and the data is automatically saved in FITS format. However, time-series operation of the CCD can only be accomplished in the "focus frames" mode which precludes the use of the smaller built-in ST-4 CCD as an autoguider.

The observer begins the session by slewing the telescope to the desired search field. *CCDSoft* is then instructed as to the number of integrations desired, duration of those integrations (i.e., integration time), and the dead time between each integration. Throughout the night, the telescope tracks on the search field and the software automatically acquires and saves the time-series data without the need for operator intervention. This mode of operation, while a bit risky in terms of incoming weather fronts, is necessary in order to allow undergraduates to participate in the project and still attend classes.

3.3. Data Reduction and Analysis

At the end of each observing session, the raw data, in FITS format, is archived on the PC to a writable CD. This disk serves as a transport medium from the data acquisition PC to the data reduction and analysis UNIX workstation. It also serves as a permanent storage archive of the raw data in the event that re-reduction is necessary. Each CD can store one to two nights' worth of data. In Figure 2 we show a flow chart of our data path from observations to identification of extra-solar planet transits.

The photometric data is reduced (see Da Costa 1992) on a UNIX-based workstation running the IRAF data reduction software. Scripts in IRAF have been written to perform the basic image calibrations (e.g. flat fielding), locate the X, Y positions of all the stars on the frames (using DAOfind), and to send this coordinate list to APPHOT which performs aperture photometry in a range of aperture radii for each star on each frame and writes the output to the UNIX disk.

These photometric output files are then fed into custom programs (written in FORTRAN) which perform ensemble differential photometry (e.g., Honeycutt 1992) on the entire dataset and assigns proper errors to each measurement (see

Howell et al. 1988). The comparison ensemble is composed of bright stars that appear on every frame. The technique used is described in detail by Howell et al. (1988), Gilliland & Brown (1992), Honeycutt (1992), and Gilliland et al. (1993). The differential photometry results are then assembled into time-series light curves which show the behavior of each star throughout each night. Statistical analysis is performed on each light curve and the database of generally over 1000 stars is searched for variability above a user defined threshold (Howell et al. 1988). When a light curve's variability exceeds this threshold, it is flagged and placed into various bins of different types of variability (for example, rapid, slow, variation only below the mean, etc.) Those light curves which are possibly due to a transit are then manually inspected by one of our group for validity and any tell-tale signatures of an extra-solar planet. The final process of light curve separation saves the staff the job of manually inspecting many thousands of datasets, the majority of which will show no variability or variability unrelated to transits.

4. Initial Results

We have observed several search fields and tested the system under various observing conditions. As a representative example we present data and discuss results from a subsection of a search field centered at $\alpha = 7^h31^m30^s$, $\delta = +9°56'24''$ ($l^{II} = 208°$ $b^{II} = +13°$; J2000) containing 78 stars in the magnitude range 10.4 to 15.5. This data set consists of over fifty 180 second exposures taken every 5 minutes (exposure + readout time) for a duration of 4.6 hours.

At our current rooftop site we have found that the typical stellar FWHM is ∼5″ and undergoes some frame-to-frame variation. This point spread function is a combination of the seeing, tracking errors, and the optical quality of the telescope. We also find significant variations in the PSF of stars with position on a single image that can be attributed to coma. The poor seeing has both advantages and drawbacks. Broad stellar profiles allow us to observe brighter stars since many pixels can be well exposed without saturation. The brightest stars are those for which the highest photometric precision is attained and sensitivity to transit events is likely to be high. Large PSFs also provide well sampled images leaving us free of worrisome effects related to undersampled PSFs (Buonanno & Iannicola 1989, Howell et. al., 1996). The drawback to larger PSFs is that the fainter stars are noisier than they would be with better seeing.

Upon inspecting the light curves produced by ensemble differential photometry performed on a night of observations we unexpectedly found that the majority of the bright stars exhibit systematic "variations". The fluxes appear to increase or decrease with time (relative to the calibration ensemble) and, in some cases, by several percent in intensity over the course of a night. Furthermore, the character of the variations is correlated with the location of the star in the field of view. The same trends may exist in fainter stars, but become masked by the noisier light curves. We believe these effects are artifacts produced by variations in photometric calibration across the field of view which have become evident in our data primarily due to poor tracking (the telescope is not guided) and PSF changes with time and location on the CCD. The telescope tracking suffers from both periodic errors produced by the drive gears and, most significantly, a systematic drift attributable to poor polar alignment and mount stability. Unfortunately improving on the polar alignment is difficult given the quality of the telescope mount. Because of these problems, the image moves across the CCD during the night and photometric calibration is affected by any systematic effects that depend on the position on the CCD. One source of these calibration problems is the variation in the point spread functions (PSFs) of stars across the CCD due to coma. Another problem may be differences between our twilight sky flats and the illumination of the actual data frames. Note that if the tracking were good, our photometric accuracy would be still be affected by these calibration problems, but the photometric precision (our main concern here) should reach its theoretical limit.

The prototype telescope system equipment is obviously the cause of most or all of these calibration problems so we expect that the problems would be eliminated or reduced to an insignificant level by the use of better equipment. The most significant improvement would likely result from better tracking, a more stable mount, and a better CCD. Note that the "problems" and level of error we are concerned about are generally not an issue for imaging or routine photometry but are of great concern here since we are working at ultra-high precisions.

Although our current hardware may ultimately limit the photometric precision, some accommodations have been made to the data reduction process that significantly reduce these problems and improve the photometric precision to the theoretical S/N limit. In response to the variations in the PSF with position, we are performing photometry using larger apertures than would be selected on the basis of an optimal S/N calculation (Howell 1992). Because larger apertures guarantee inclusion of the total flux, the effect of differences between the PSF of each star and the comparison ensemble is reduced. Another approach we have taken is to confine the entire data reduction process to small subsections of the CCD with the assumption that differential photometry will eliminate the systematic errors that neighboring stars (those in the same CCD subsection) appear to have in common.

The light curves of three stars taken from our example data set are shown in Fig. 3. The magnitudes assigned to these stars were derived by comparison to stars in the same field with published V-band magnitudes from Henden & Honeycutt (1995). We assign the comparison stars the same magnitudes in the bandpass ($\sim$V+R) of our observations. This allows us to assign a rough magnitude to each star. The stars in Figure 3 are chosen to represent three different magnitudes and different signal-to-noise conditions. They do not represent our brightest ($V \sim 9$) or faintest ($V \sim 15$) photometric datasets. Since we don't have spectroscopic or color information for any of these stars, their spectral and luminosity class is currently unknown. To produce our example data we have confined the field of view by considering stars only in a $18' \times 18'$ subsection of the CCD. The photometric extraction aperture diameter is $23''$ ($4.6 \times$FWHM) and the magnitudes listed in the plots are the approximate V magnitude of the star. For 180 sec exposures, our system bright limit is near V=9, above which saturation occurs. No statistically significant variations are observed in the sample light curves shown in Fig. 3 providing an idea of our obtainable photometric precisions. The error bars on each point represent theoretical uncertainties calculated using the complete S/N formula given in §2, properly applied to differential measurements (Howell et al. 1988; Honeycutt 1992). The standard deviation actually *observed* for each entire light curve is labeled on the plot. There is good agreement between the observed and theoretical standard deviations for each light curve. This indicates that the photometric precision is almost entirely limited only by the various factors appearing in the S/N equation (Howell & Merline 1995). The observed standard deviation of the light curves in Figure 3 is the best estimate of our observational precision (assuming the stars are nonvariable), but the theoretical uncertainties are useful since they can be calculated for each individual data point and provide a predictive base level of variation which we should expect. Notice that the measurement uncertainties increase with time in this data set as the atmospheric extinction increased due to increasing airmass.

In order to test our software routines searching for possible transit signatures, as well as to convince ourselves that we could unambiguously detect transit level photometric excursions, we will look at a sample case. Suppose the top light curve in Figure 3 represents a G dwarf and a Jupiter-sized body transits, causing a 1% diminution in light. To simulate a light curve of this type of transit, we consider a duration of 2 hours similar to that expected for a 51 Peg-like system. In the top panel of Figure 4 we show the light curve for the brightest star of Figure 3, but with 1% of its flux subtracted for two hours. In the bottom panel we show the same data including the simulated transit, but with the data points binned and averaged in sets of 3. The transit event is detectable as a candidate since multiple magnitude measurements are made during the transit, however the deviation of individual measurements from the mean magnitude of the star outside of the transit is not much greater than the uncertainties for this case. For brighter stars ($V = 9 - 10$) or later spectral types, the increased precision and larger $\Delta I/I$ will be of benefit. As discussed in §2, we would likely observe multiple transits for confirmation of any 51 Peg-like systems and could co-add multiple transits for higher S/N and further information on ingress and egress.

If most stars in our field are nonvariable, including the calibration ensemble, and assuming no other sources of systematic error occur in the photometric measurements, the light curve of each star will be constant and we can check our photometric precision as a function of magnitude by measuring the variance in the light curves for each star in the entire data set. In Figure 5 we plot the logarithm of the standard deviation (in mags) as a function of measured magnitude for all of the light curves in our example data set. Again, magnitudes have been assigned to these stars based on a comparison to the V-band standards in our field as described above. The expected curve representing the theoretical photometric precision predicted for our sky brightness, other sources of noise, and star counts is also shown. Most of the data points fall along or slightly below the curve as expected.

5. Discussion

The photometric search discussed above will be limited to possible ESP transit detections for large-sized bodies within a magnitude range of roughly V=9-12 (see Fig. 5). The brighter limit is set by saturation of the CCD, while the fainter is due to the minimum S/N needed to detect at least a Jupiter-sized body. For Jupiters around K and M stars, we will be able to use stars down to $\sim$ 13.5 magnitudes. Our best photometric errors are near 0.004 at 9th, to 0.009 at 13th. At a galactic latitude of say 30° there are about 100 stars per square degree between 9th and 13th mag (Bahcall and Soneira 1980). Covering approximately one-quarter of a square degree per CCD frame, we will observe about 25 stars bright enough for which we could possibly detect ESP transits. If 70% of the stars are solar-type or later dwarfs (30% will be earlier types and/or giants [Allen 1963]), we will have 18 suitable targets in each field-of-view. If we assume that each of these stars has one (and only one) planet like 51 Peg (ie., with a 4.2 day orbital period), approximately 10% ($\sim$ 2) of these planets' orbits will lie close enough to our line-of-sight ($\lesssim$ 5°) that they will transit the star. If the same field-of-view is observed on 50% of the nights for two months (25% time coverage), then we would typically observe $\sim$ 6 transits in total among these $\sim$ 2 planets. It is therefore likely that we would detect all of the transiting 51 Peg-like planets in our search field. Detecting transits by planets in Jupiter-like orbits is more difficult since the orbital period is much longer and the possible inclinations leading to a transit are more restricted. Of course not all late-type dwarfs have planets like 51 Peg. Using current discoveries and caution about small number statistics, $\sim$ 30% of late-type dwarfs harboring planets will be similar to 51 Peg. Thus, by covering multiple fields there is a good chance that our current system will either detect a number of planetary transits or set significant limits on the number of planets having certain orbital characteristics. Improving the detection probabilities is accomplished by increasing the number of suitable target stars. This can most easily occur by increasing the survey area (wide field imaging, multiple fields), increasing the usable magnitude range (larger telescope, better CCD), or both. We hope to obtain both of these improvements in the near future.

A ground-based wide-field photometric search for extra-solar planet transits represents an alternative and complement to other search techniques. Transit-based detections, although probably favoring the detection of giant planets in short-period orbits similar to radial velocity detection methods, allow for the detection of a wider range of possible planet-star combinations. With this method, planets orbiting stars of a wide range in spectral class (F and later type dwarfs) are detectable. Photometry can also be used to detect a wide range in planet size (or mass). While spectroscopic and astrometric searches are presently confined to detecting massive planets, searches for transits offer the possibility of detecting terrestrial-sized planets. We note that terrestrial-sized planets would not suffer from confusion with brown dwarfs. Another important consideration favoring ground-based transit searches of the type described here is the very low cost (e.g., see Table 2).

Perhaps the most significant contribution that photometry can make to the search for extra-solar planets is that a large number of stars can be monitored, and once candidate systems are identified, detailed follow-up studies (e.g., spectroscopy) can and should be carried out. If feasible, high resolution spectroscopy should be used to derive the candidate planet's mass and orbit. The stars sampled by our survey are too faint for most current ground-based radial velocity search programs, however good transit candidates would be appropriate for follow-up spectroscopy with the Keck or other large telescopes.

Observations of transiting extra-solar planets provide two important pieces of information generally lacking in systems discovered through radial velocity measurements: orbital inclination and planet size. Knowing the inclination eliminates the associated ambiguity associated with planetary mass estimates. A candidate planet's size may be combined with its mass to determine its density, a vital piece of information towards understanding these objects. For some systems, photometric information alone may be complicated by confusion between planets and white dwarfs or brown dwarfs. This problem may be eliminated if the mass is determined and found to be compatible only with that of a planet. Even so, the meridian along which transits occur and identifying partial, grazing transits may remain a complication.

It will be important to conduct certain follow-up observations of all transit candidate stars. The spectral type and luminosity class of the star are necessary to estimate a planet size from the depth of a transit. The stars in our survey, as well as those that COROT and KEPLER will observe, are bright enough that classification spectroscopy can be carried out with a small or medium size telescope.

Planetary transits, when the precisions and methodology are further developed, can reveal many details about planets and their stars. Saumon et al. (1996) has shown that determining the radius of any detected extra-solar planet will be of prime importance. Due to the equations of state governing gas giants and rocky planets, planets of mass up to 1-3 times that of Jupiter, if made of silicates, will have radii of only 0.3-0.5 R_J. Jupiter-like gaseous planets, even if more massive, will have radii approximately equal to or only slightly larger than that of Jupiter. Measurements of planetary radii to values of 1-3% would allow estimates of the age, surface temperature, luminosity, and composition, once the orbital parameters are known. Additional information may be found from modeling well sampled, high precision transit light curves. Transits of planets with large satellites should show an additional light decrease when the satellite itself transits. The effect of planetary ring systems would be similar. Transiting planets will also be eclipsed by their stars and experience different phases of illumination. Photometry of the system at different phases and as the eclipse occurs can tell us about the planet's albedo. During a transit minimum spatial variations in the star's surface brightness (limb brightening or starspots) would be revealed and monitoring observations would show temporal changes in the star's atmosphere. Spectroscopy during a transit could allow radial velocities to be measured such that the planet's path across the star and the angle of the star's rotation could be determined.

An important goal of our search program is to discover limitations and develop techniques for ultra-high precision photometry. At the level of photometric precision we expect to achieve (0.1% or better), previously unidentified sources of error will be discovered. At this level, CCD manufacturing techniques and software algorithms become important. We will explore the detailed issues that could affect our photometry including aperture centering (Stone 1989), the treatment of partial pixels (Merline & Howell 1995), unusual stellar PSFs (Diego 1985, Howell et al., 1996), different types of stellar photometry (e.g., PSF fits vs. aperture photometry), intra-pixel flat field differences (especially for large image scales;Jorden et. al., 1994), non-linearity of the CCD (Massey & Jacoby 1992), and flat-fielding procedures. Using defocused stellar images may provide a solution to many of these effects, but as yet no quantitative study of defocusing has been performed.

In addition, we will study means of distinguishing between transits and other stellar phenomena (flares, starspots, other [known and unknown] variability etc.). Long-term monitoring of all stellar types at these photometric precisions is unprecedented. Our work will be important groundwork for future costly satellite missions. For example, a large sunspot group moving across the sun may decrease the total solar flux by up to 2%. This lasts for the entire rotation and is much larger than the transit signals we are looking for. How do similar events affect other stars? Solar observations show that color effects may be useful to disentangle transits from atmospheric activity. However, further study on solar color reveals that sunspots are red but faculae are blue. Thus, at times their effects on color can cancel, or produce a redder or bluer star. Also, using the sun as a baseline may be dangerous as it is well known that the sun brightens during solar maximum, while other active stars seem to drop in brightness with increased spot activity.

Ground-based searches for extra-solar planet transits are feasible and can be carried out with small telescopes. Transit-based searches offer certain advantages over other planet detection techniques in use today. Additionally, photometry combined with radial velocity discoveries would be a powerful combination for confirmation of, and obtaining physical parameters for planetary candidates. The photometric precision achievable with the system we have described is in fact capable of detecting terrestrial-sized planets transiting M dwarfs. There are, however, many complications that will require further attention to fully utilize this method. Success will require developing new observational and data reduction and analysis techniques as well as building a knowledge base of low-level stellar variability from long-term photometric monitoring of field stars. Ground-based studies like ours represent an important step towards future space-based missions while in the meantime providing a good chance of discovering extra-solar planets.

Acknowledgments. We wish to acknowledge funding and support received from the NASA Origins Initiative via grant NAG5-6492, and from the Wyoming Space Grant Consortium funded by NASA. Harley Thronson and Paul Johnson are to be thanked for their personal support of this idea.

References

Allen, C. W., 1963, *Astrophysical Quantities*, Althone Press.
Bahcall, J., & Soneira, R., 1980, ApJS, 44, 73.
Borucki, W., & Summers, A., 1984, Icarus, 58, 121.
Buonanno, R., & Iannicola, G., 1989, APSP, 101, 294.
Cochran, W., Hatzes, A., Butler, P., & Marcy, G., 1997, ApJ, 483, 457
Da Costa, G., 1992, in "Astronomical CCD Observing and Reduction Techniques", Ed. S. Howell, ASP Conf. Series, Vol. 23, p 90
Diego, F., 1985, PASP, 97, 1209
Giampapa, M., Craine, E., & Hott, D., 1995, Icarus, 118, 199
Gilliland, R., & Brown, T. 1992, PASP, 104, 582
Gilliland, R. et al. 1993, AJ, 106, 2441
Henden, A. A., & Honeycutt, R. K., 1995, PASP, 107, 324.
Honeycutt, R. K., 1992, PASP, 104, 435.

Howe, H., 1909, "*Elements of Descriptive Astronomy*", Silver Burdett & Co. New York, p 252

Howell, S. B., Mitchell, K. J., & Warnock, A., III, 1988, AJ, 95, 247.

Howell, S. B., 1989, PASP, 101, 616.

Howell, S. B., & Merline, W., 1995, in "New Developments in Array Technology and Applications", Eds., A. G. Davis Philip, K. Janes, & A. Upgren, p371

Merline, W., & Howell, S. B., 1995, Exp. Ast., 6, 163.

Howell, S. B., Koehn, B., Bowell, E. L. G., & Hoffman, M., 1996, AJ, 112, 1302

King, I., 1971, PASP, 83, 199

Lardner, D., 1858, "*Hand-Books of Natural Philosophy and Astronomy*", Third Course, Blanchard and Lea, Philadelphia, Chp XXVII

Marcy, G., Butler, P., Williams, E., Bildstom, L., Graham, J., Ghez, A., & Jernigan, G., 1997, ApJ, 481, 926

Marcy, G., & Butler, P., 1997, Bull. AAS, 191

Marcy, G., & Butler, P., 1998, Sky & Telescope, 95, No.3, p 30

Rosenblatt, F., 1971, Icarus, 14, 71

Saumon, D., Hubbard, W. B., Burrows, A., Guillot, T., Lunine, J. I., & Chabrier, G., 1996, 460, 993.

Schneider, J., & Chevreton, M., 1990, A&A, 232, 251

Stone, R. C., 1989, AJ, 97, 1227

Young, A., 1974, ApJ, 189, 587

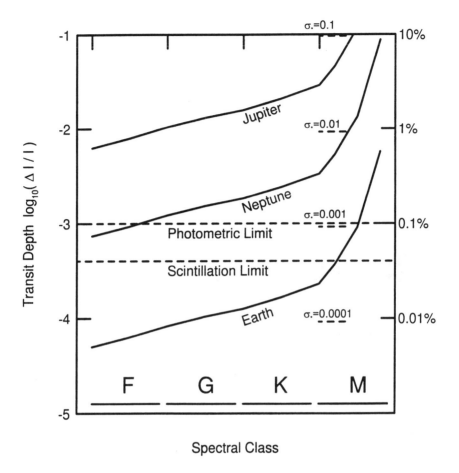

Figure 1. This diagram shows the limit of photometric precision needed to detect various sized bodies in orbit around F, G, K, and M dwarfs. The three curved lines represent planets of size equal to Jupiter, Neptune, and the Earth. For ground-based telescopes, Jupiter-sized bodies are easily detected around all of the spectral types plotted here, Neptune-sized bodies are likely detectable for G-M stars, and *earth-sized bodies may be detectable* for late M stars. No other current or near-term planned search system can do all this. The horizontal line labeled "photometric limit" is the best theoretical possible for our current, "off-the-shelf" photometric search system. This value is 0.001 magnitudes, while our best actual data from our initial 6 month assembly and testing phase, is 0.004 magnitudes or 0.4% photometric precision. Scintillation in the earth's atmosphere is the ultimate limit on ground-based photometric precision and it has a value of 0.0004 magnitudes (or 0.04% photometric precision) at 1 airmass and moderate altitude.

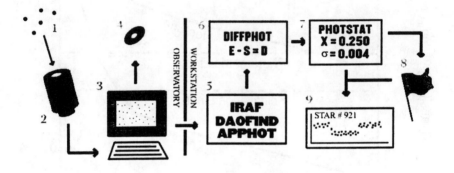

Figure 2. A flow chart of our data path from observations to identification of potential extra-solar planet transits. We obtain images towards a chosen search field every few minutes on each clear night over a period of a few months (steps 1 and 2). A PC is used to control the CCD and save each image on a hard drive (3). We archive the images on writable CDs at the end of the night (4) and transfer the data to a UNIX workstation for data reduction and analysis (5). We then run IRAF scripts on the workstation to locate and perform aperture photometry on all of the stars in the search field (6). Next, the instrumental magnitudes produced by IRAF are flux-calibrated with Fortran codes using ensemble differential photometry techniques. Finally (7), statistical tests are run on the light curve of each star to search for variability and flag candidate extra-solar planet transits (8 and 9) which can be inspected manually and, if promising, followed up with other observations and analysis.

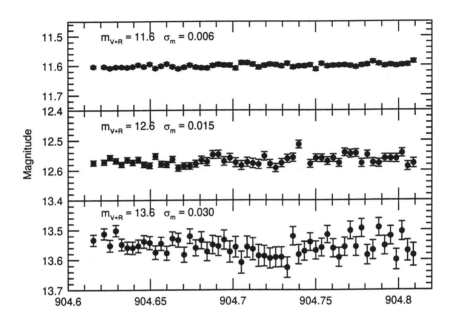

Figure 3. Light curves for three stars at different representative magnitudes observed over a period of approximately 4.5 hours. The magnitude (approximately V+R-band, but measured in white light; see text) of each star and the standard deviation of the entire light curve σ_m are labeled. The error bars represent error predictions for each measurement based on total counts and noise from the sky background, read noise, and dark current. The standard deviation of the entire light curve is consistent with the theoretically calculated error bars (σ_*) thus no indications of variability are seen in these three light curves. For stars at our bright limit of V=9, we can obtain precisions of 0.004 mags.

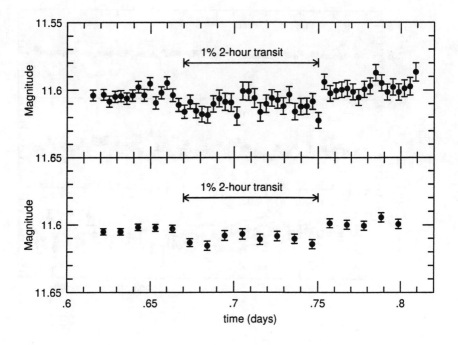

Figure 4. A *simulated* transit added to the light curve of the brightest (nonvariable) star that is shown in Figure 3. The top panel shows the same light curve as in Figure 3, but with the flux artificially diminished by 1% for a period of 2 hours as indicated by the line drawn over the data points. The bottom panel shows the same light curve plus the simulated transit, now with the data points binned and averaged by sets of 3.

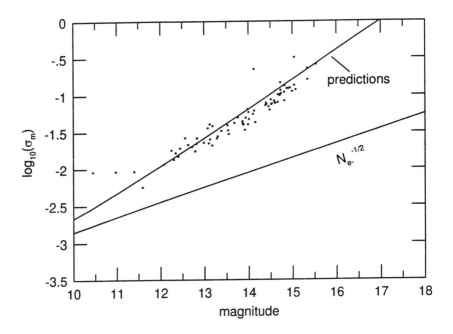

Figure 5. A plot of the logarithm of the standard deviations of a set of simultaneous light curves versus the mean magnitude of each light curve. The magnitudes are approximately V+R-band, but measured in white light (see text). Two lines on the plot show the expected deviations derived from the theoretical S/N formulas discussed in §2. The top line (labeled "predictions") represents the expected result when all sources of noise are accounted for. The bottom line represents the ideal case where the only source of noise is Poisson fluctuations in the counts from the star itself. Most of the observed σ_m follow the predictions, however, σ_m for a few of the brightest stars exceed the predictions. This is due to the position-dependent calibration problems discussed in §4.

CCDPHOT Photometry of Extremely Metal-Poor Stars

B. J. Anthony-Twarog

University of Kansas

T. C. Beers and S. L. Hawley

Michigan State University

A. Sarajedini

San Francisco State University

B. A. Twarog

University of Kansas

Abstract. We have used an extended version of the Strömgren photometric system to study candidates in a search for extremely metal-poor stars.

1. Introduction

Like forensic scientists or archaeologists, astronomers scavenge among meager clues to reconstruct a story played out eons ago. While some astrophysicists attempt to model the processes that lead to the formation of galaxies in the Universe, many believe that important evidence survives among the Galaxy's oldest stars.

Extremely metal-poor stars in the solar neighborhood are relics of the Galaxy's earliest episodes of star formation. They are rare stars in our neck of the woods; about one in a thousand stars in the field is a halo "interloper" speeding through the younger, gas-rich galactic disk. Nevertheless, we expend considerable effort to find and study these extremely metal poor stars in order to understand the chemical and stellar evolution processes at work 10 billion years ago.

The first task is to develop a strategy to identify the Galaxy's oldest stars in our midst. One popular strategy has been to exploit the large velocities relative to the sun's motion around the center of the Galaxy. A star with a large space motion is nearly certain to be a halo star, but the converse isn't necessarily true and this approach has led to some biases in past surveys, overlooking stars with low chemical abundance and coincidentally slow motion relative to ours. A better and more direct method is to use the chemical underabundance of the stars themselves as a survey criterion. This was the motivation behind the surveys of Beers and co-workers (Beers, Preston & Shectman 1985). Their survey technique used an interference filter to isolate the spectral region containing the strong lines of ionized calcium in the near ultraviolet, one of the few lines detectable in low-metallicity stars. This filtering was imposed on wide field images taken with a Schmidt camera with a dispersing prism. Candidate stars with very low abundance could be identified by their weak calcium line features. In their initial survey of southern hemisphere plates, 2000 candidate stars were identified. Of these, 450 stars have been confirmed as stars with [Fe/H] ≤ -2.0, 70 of them lower than -3.0 and only a dozen or so with [Fe/H] lower than -4.0.

Beers has obtained northern hemisphere prism plates to survey in a similar manner, and has assembled teams to assess as many of the candidate stars as possible. We (Beers, Sarajedini, Anthony-Twarog, Twarog and Hawley) were granted status as a Key Project for several years at Kitt Peak National Observatory, to use a two-fold approach combining spectroscopy and photometry to assess the nature of the candidate metal-poor stars. This report summarizes the results of our photometric surveys among these metal-poor candidate stars, obtained at the 0.9m telescope in 6 runs between March 1995 and October 1997.

2. The Key Project Study

Successful application of photometric techniques to stellar population studies depends on a thoughtful match of the photometric system's capabilities to the class of stars under study. A number of researchers have worked for some years to augment and extend the utility of the Strömgren photometric system to fainter, cooler and older stars. As originally introduced by Strömgren (1966), this intermediate band system employs four filters - y, b, v and u - with bandpasses approximately 200 Å wide. From these four filters, the following indices are constructed in addition to $(b - y)$, the basic color temperature index analogous to $(B - V)$:

$$m_1 = (v - b) - (b - y)$$
$$c_1 = (u - v) - (v - b)$$

The m_1 index measures the difference between two color ratios due to blanketing by metal lines in the v bandpass. The c_1 index measures surface gravity by quantifying the the difference in color slope on either side of the Balmer jump.

In a series of papers, Crawford developed calibrations of these indices as well as an Hβ index for absolute magnitude and intrinsic colors of B (Crawford 1978), A (Crawford 1979) and F stars (Crawford 1975). Extension to cooler stars has long been a feature of Eggen's work with a slightly modified form of the Strömgren system (Eggen 1976, 1978). We have built upon the early observations of Bond (1980) of metal poor red giants (Anthony-Twarog & Twarog 1994) while a slightly different calibration for cooler stars is embodied in the extensive work by Olsen (1993).

Application of the Strömgren system to fainter stars has been facilitated by the availability of CCD's, particularly with enhanced ultraviolet sensitivity. Extension of the system to more metal-poor stars, however, hit a more fundamental snag. The m_1 index, because it measures the effect of rather weak metal lines in the v bandpass, loses sensitivity at abundances typical of halo stars, a problem more commonly associated with broad-band systems at even higher

[Fe/H]. We developed and calibrated a new index, hk, that replaces the v filter's reliance on weak iron lines with a filter, Ca, that measures the depth of the ionized calcium H and K lines in the near ultraviolet. The hk index is defined analogously to m_1 as $hk = (Ca - b) - (b - y)$ (Anthony-Twarog et al. 1991; Twarog & Anthony-Twarog 1995). As hoped, the hk index is two to three times more sensitive to changes in [Fe/H] for cooler and more metal-poor stars than m_1.

3. CCDPHOT

The CCDPHOT instrument is a software adaptation of an imaging CCD system, designed to mimic the speed and simplicity of a photoelectric photometer. This is accomplished by utilizing a small subraster of the CCD chip, automating the bias and flat-field corrections, and setting up immediate magnitude calculations within defined apertures centered on the star nearest the chip center.

Although this particular instrument has been retired from use at Kitt Peak National Observatory, the principles involved are not so different from the actions that most photometrists would take to ensure uniform data acquisition from a photoelectric system. As a photometrist would select an aperture several times larger than the seeing disk of stars and would retain that aperture throughout the night, CCDPHOT "selects" aperture radii within which magnitudes can be calculated. The small chip size makes it possible to cycle through filters quickly and repeat observations for each filter for each star. With some slight improvement over photoelectric observations, the sky is measured simultaneously with each filter, calculated from a specified annulus outside the stellar measurement aperture. After a night's observing, CCDPHOT presents the sleepy user with a list of magnitudes, one for each filter and star.

4. The Semi-Software

Because we were committed to a survey project, which implied single observations for as many stars as possible, we did not plan to repeat observations for more than a few stars from night to night or run to run. To give some indication of the repeatability of separate flux measurements, we elected to use the less condensed output of the CCDPHOT system which retains each recorded flux measurement for each filter. In some cases, all meager fluxes for a faint star were summed for the magnitude calculation, a somewhat safer procedure than averaging noisy magnitudes. This also supplied us with magnitude errors based on the deviation in measurements within a long measurement cycle, not just photon statistics.

We have followed a practice of generating a master file of instrumental photometric indices for a given run by transforming indices from each night to a standard night and then averaging to produce higher precision instrumental indices. All the nights must, of course, be scrupulously photometric and there needs to be substantial overlap between nights, in most cases supplied by the standard stars. We extended this practice and transformed runs to a master instrumental list. Some of these transformations entail color terms which accommodate the secular changes in filters and the replacement of detectors. We then compute the transformation to the standard system for the entire merged set, using the merged values for the standard stars.

To maintain the strongest connection to our prior studies of red giants, we transformed m_1 and c_1 indices for evolved stars to values from Anthony-Twarog & Twarog (1994), equivalent to the Bond (1980) system. The V magnitudes, $(b-y)$ colors and hk indices for all stars are tied to the Twarog & Anthony-Twarog (1995) catalog. Metallicity and surface gravity indices for the bluer stars and cooler dwarfs are transformed to the system defined by Schuster and Nissen's observations of metal poor dwarfs. The standard deviations characterizing the transformation of V, $(b-y)$ and hk to the standard system are 0.012, 0.009, and 0.016; the comparable statistics for m_1 and c_1 are 0.019 and 0.020.

Over the six-semester lifetime of the Key Project observations, we did manage to obtain multiple observations of approximately 100 program stars; these give a separate indication of the repeatability of our photometric indices from one run to another. Excluding a few of the worst performers, those deviations for V, $(b-y)$, m_1, c_1 and hk are 0.029, 0.014, 0.014, 0.021 and 0.030 for stars that reach as faint as $V = 15$.

5. The Results

Figure 1 presents the final product of these observations, the photometric plane showing hk and $(b-y)$ indices for 571 program stars between $V \sim 10$ and 16. Superimposed on the figure are the recently recalibrated isometallicity lines for this photometric plane, separating stars with [Fe/H] values between -0.5 and -3.4; stars above this last isometallicity curve may be of even lower abundances, although the effects of reddening have not been corrected for in this plane and would move stars to the right. In the coming months, we will attempt reddening corrections and a closer look at the population characteristics of the most intriguing stars in the sample.

It is important to emphasize that a program of this type would not have been feasible without a system like CCDPHOT. Extensive survey projects with large format CCD's can generate thousands of stars which require follow-up observations; the candidate list of extremely metal-deficient stars numbered more than 5000. If, as was the case here, the candidates are spread over large areas of the sky, individual observations of each candidate may be the only means of obtaining the required information. Traditional single-star photometry with a CCD detector and simultaneous star-sky observations, as realized by CCDPHOT, permits small telescope users ($\sim$1-m class) to attack problems that required 4-m class telescopes a decade ago. The need for the information provided by intermediate and narrow-band photometry has not disappeared. If anything, the tidal wave of information generated by recent and planned surveys makes selective follow-ups even more imperative. CCDPHOT demonstrated the the potential that can be achieved by a merger of contemporary detector and analysis systems with the design and philosophy of more traditional single-star photometric techniques.

Acknowledgments. We have enjoyed mining this excellent database, although we regrettably only scratched the surface of the entire sample of northern BPS stars. Our thanks to Tim Beers for inviting our participation on this

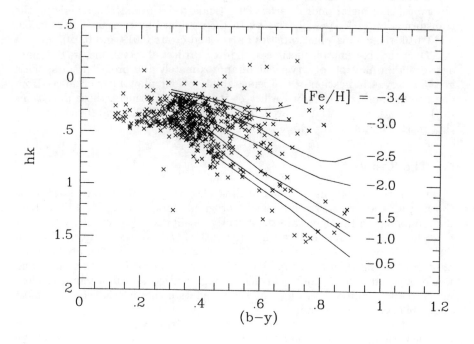

Figure 1. $hk,(b-y)$ indices for 570 metal-poor candidate stars. Labeled isometallicity lines for [Fe/H] = -0.5 to =3.4 are superposed.

project. Ata Sarajedini took a lion's share of the data while on staff at Kitt Peak.

We owe a great deal to the staff of Kitt Peak National Observatory, in particular to Ed Carder for his assistance in setting up so many nights, and to the developers of CCDPHOT, a fine marriage of instrument and software.

Some of our travel (BAT and BJAT) has been supported by the University of Kansas Department of Physics and Astronomy, some by funds from the National Science Foundation through its EPSCoR grant to the state universities of Kansas.

References

Anthony-Twarog, B. J., Twarog, B. A., Laird, J. B., & Payne, D. 1991, AJ, 101, 1902

Anthony-Twarog, B. J. & Twarog, B. A. 1994, AJ, 107, 1577

Beers, T. C., Preston, G. W. & Shectman, S. A. 1985, AJ, 90, 2089

Bond, H. E. 1980, ApJS, 44, 517

Crawford, D. L. 1975, AJ, 80, 955

Crawford, D. L. 1978, AJ, 83, 48

Crawford, D. L. 1979, AJ, 84, 858

Eggen, O. J. 1976, PASP, 88, 732

Eggen, O. J. 1978, ApJ, 221, 881

Olsen, E. H. 1993, A&AS, 102, 89

Strömgren, B. 1966, ARA&A, 4, 433

Twarog, B. A. & Anthony-Twarog, B. J. 1995, AJ, 109, 2828

Precise Photometry of the GSC 3493_742 Field

R. M. Robb and R. Greimel

Department of Physics and Astronomy, University of Victoria, Victoria, BC, CANADA V8W 3P6

Abstract. Precise photometric observations of the Ca H&K emission line star GSC 3492_742 were made with the 0.5m telescope of the Climenhaga Observatory. GSC 3492_742 was found to be variable in brightness with a period of about 0.81 days and an amplitude of about 0.02 magnitudes. Flares were observed implying that it is a rotating spotted star of the BY Dra type. Two additional variable stars were serendipitously discovered in the field: GSC 3493_1158 is a detached eclipsing binary with a period of about 1.23 days, and GSC 3493_1097, a W UMa type with a period of 0.32229 days.

1. Introduction

Precise photometry using CCD's has been done at major observatories for the past twenty years, and is now within the financial and time constraints of the smaller University observatories. We have attained millimagnitude differential photometry with our 0.5 meter automated telescope and use the system for a number of different projects. Smaller telescopes have the great advantage that projects requiring quick response, specific times, or long time spans can be accomplished. We find that at the millimagnitude level of precision a number of background stars are discovered to be of interest.

One of our current projects is the identification of stars with active regions. A search by Beers et al. (1994) of part of the sky for stars with Ca H&K emission yielded the star BPSBS16029-24 = GSC 3493_742. This star has been classified by Stephenson (1986) to be a K7 dwarf and UBVRI measurements are given by Weis (1991). The GSC identifications and J2000 positions from the Guide Star Catalog (Jenkner et al. 1990) for all the measured stars in the field are given in Table 1.

2. Observations

The telescope used was a F/10.5-0.50 meter Cassegrain situated on top of the physics and astronomy building. A STAR I CCD camera manufactured by Photometrics with 576 × 384 23 micron pixels gave a plate scale of 0.86 arc seconds per pixel. The camera was cooled by a thermoelectric refrigerator running without temperature regulation to a temperature of about -55C. This temperature gave a dark current of approximately an electron per second per pixel, which was equal to the contribution from our rather bright sky. The gain was 40.8

electrons per ADU and with the 12-bit Analog to Digital converter the read noise was 22.4 electrons. The automatic data gathering system was described in Robb and Honkanen (1992). Briefly, a personal computer (PC) communicates with the elements of the system through an interface box. The components of the system are the dome azimuth, dome closing, telescope motions, filter positions, camera shutter, CCD controller, rain sensors, and UT/LST clock. The astronomer must open the dome, find the star of interest, designate the guide star, set the filters and exposure times. To guide the telescope, the PC finds the brightest pixel in the guide box subrastered from the science frame and moves the telescope to keep that star on the same pixel. The telescope tracking is adequate for exposures of a few minutes and long exposures are precluded by our bright sky. The PC will then keep taking pictures of that star field until dawn, clouds, rain or high airmass force a halt to operations. At that time the PC will close the dome and turn off the drive. Thus the "observer" was NOT in the dome for almost all of these observations.

3. Analysis

Using IRAF [1] routines CCDPROC and PHOT the frames were de-biased and flat fielded, and the magnitudes were found from 6 pixel aperture photometry after using the Centroid centering option. The sky ANNULUS used was 15 pixels and the DANNULUS was 25 pixels for an area of 4300 pixels and the mode of the sky values was approximately 50 ADU's. Our seeing is not particularly good so the FWHM of the stars was approximately 5 pixels. This is not of great consequence since the stars we are observing are generally quite bright. The filter used was the Cousins R band and the exposure time was 133 seconds. The brightest star had a peak brightness of 1600 ADU and the IRAF MERR estimate was 0.001. The faintest star had an MERR of 0.012 magnitudes per observation.

Table 1. Stars observed in the field of GSC 3493_742

GSC No. 3493_	RA J2000.	Dec. J2000.	V Mag.	(B-V) Color	ΔR Mag.
742	$15^h58^m10^s$	+49°27'11"	11.90	1.43	$1.751 \pm .012$
712	$15^h57^m56^s$	+49°28'18"	9.81	1.04	-
1158	$15^h57^m45^s$	+49°27'55"	13.69	0.49	$4.136 \pm .072$
003	$15^h57^m54^s$	+49°31'33"	12.72	1.16	$2.833 \pm .002$
1330	$15^h58^m05^s$	+49°30'21"	11.22	0.57	$1.637 \pm .001$
1097	$15^h58^m25^s$	+49°26'51"	11.97	0.82	$2.268 \pm .016$

---

[1] IRAF is distributed by National Optical Astronomy Observatories, which is operated by the Association of Universities for Research in Astronomy, Inc., under contract to the National Science Foundation

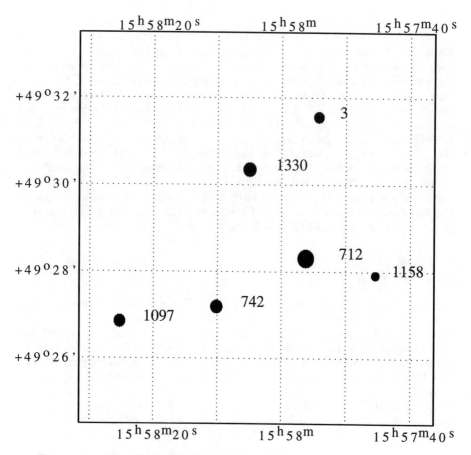

Figure 1. The GSC 3493_742 field labelled with the GSC id numbers

4. The Data

The field of GSC 3493_742 can be seen in Figure 1 labelled with the Guide Star Catalog numbers and Table 1 contains the star's J2000 positions and magnitudes.

The V magnitudes and (B-V) colors were measured differentially relative to the single star GSC 3493_712, which has V and (B-V) values (Urban et al. 1998) observed by the Hipparcos satellite (ESA 1997). The V and (B-V) values of GSC 3493_742 are consistent with those of Weis (1991). The differential R magnitudes were measured relative to GSC 3493_712. A mean and standard deviation of the differential R Magnitudes was calculated for each star for each night. The nightly means were then averaged and the standard deviation calculated to find the ΔR given in the table. The differential R magnitudes for GSC 3493_1330 are plotted in Figure 2 where the points are the average for the night and the error bars are the standard deviations of the points, not the error in the mean. A horizontal line has been added at the mean level to guide the

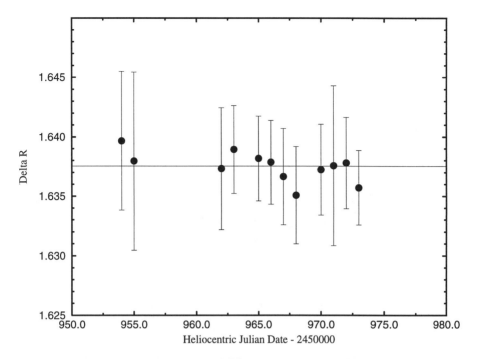

Figure 2. The nightly mean GSC 3493_1330 minus GSC 3493_712 R magnitudes.

eye. The excellent precision of the photometry is shown by the one millimagnitude standard deviation of GSC 3493_1330 and the two millimagnitude standard deviation of GSC 3493_3. Variations in the brightness of GSC 3493_1097 were obvious and the variations of GSC 3493_742 and GSC 3493_1158 became evident after a few nights.

5. GSC 3493_742

Inspection of the differential plots showed some variation on a nightly timescale, and that the period of variation was longer than half a day. A fit of the data to a sine curve revealed a minimum in χ^2 at a period of 0.81 days. A plot of the data at this period is shown in Figure 3. The amplitude of 0.02 magnitudes and shape of the light curve are consistent with a rotating spotted star. On the night of Julian Date 2450967.8 a flare with an amplitude of about 0.05 magnitudes in the R band and duration of about 15 minutes was observed. It was not of unusual power, but it stands out clearly because the scatter of the other points is only a few millimagnitudes. Because this star has Ca H&K emission and a brightness variation on the time-scale of days, we suggest the star is a spotted star similar to BY Draconis.

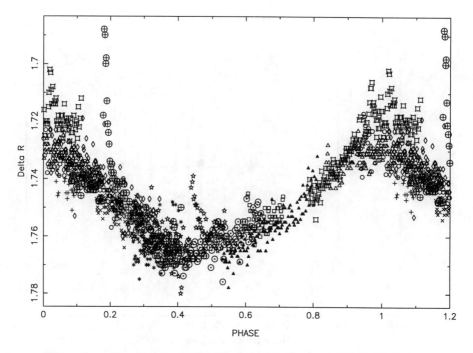

Figure 3. The Differential R Band Light Curve of GSC 3493_742

6. GSC 3493_1097

From the obvious brightness variations of GSC 3493_1097 on the second night the time period from maximum brightness to the next maximum was seen to be about 0.16 days. A sine curve was fit to the data for various frequencies around this period and the χ^2 for each trial frequency was inspected. The best fit then was 6.2 cycles/day. Twelve times of minima were found using the method of Kwee and van Woerden (1956) and are tabulated in Table 2. (HJD 2450000+)

Table 2.   Heliocentric Times of Minimum of GSC 3493_1097

955.7562	962.8495	963.8161	965.7495
966.8768	967.8437	968.8098	970.7475
970.9059	971.8746	972.8395	973.8018

A fit to these minima give the ephemeris:
    HJD of Minima = 2450954.6299(16) + 0.32229(4) × E.
with a χ^2 from the fit of 0.0020 days.

In Figure 4 the twelve nights of data are plotted according to the period 0.32229 days. The resulting light curve shows the shape of a W UMa system, Delta Scuti or a β Cephei star. The (B-V) color of 0.82 and the minima alternating in depth imply that the star is a W UMa system. The period and color

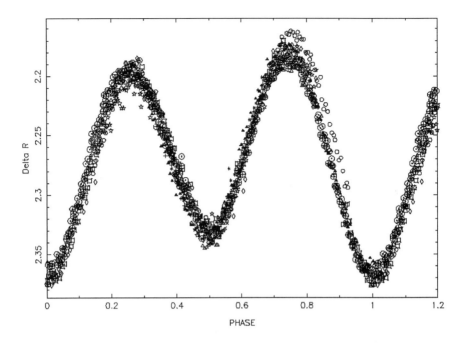

Figure 4.   The Differential R Band Light Curve of GSC 3493_1097

are consistent with the period-color relation of Rucinski (1997). The second maximum shows a difference in brightness of 0.04 magnitudes from observations separated by only four days. Fluctuations like this are to be expected in contact systems since they are often covered with large active regions.

The light curve modelling program "Binmaker2" written by Bradstreet (1993) was used to find a set of parameters which matched the light curve of GSC 3493_1097. While this is not a "solution" of the light curve it helps indicate what direction of further study would be valuable. An example model light curve (the line) and the averaged data points are shown in Figure 5. The parameters used to generate the light curve are a temperature of the large hot star of 5100K assumed from the (B-V) and since the stars are in contact the smaller star was assumed to have the same temperature. Limb darkening, reflection and gravity darkening coefficients were set to match this temperature. The mass ratio was 1.0, the fill-out factor was 0.2 and the inclination was 45°. These three parameters are correlated such that increasing the mass ratio can be compensated for by increasing the inclination and decreasing the fill-out factor. Therefore the uncertainty in the mass ratio is a factor of ~ 3, the fill-out factor is $\pm .2$ and the inclination is $\sim \pm 10^o$. A dark spot 20° in diameter and situated on the equator at a longitude of 210° was necessary to fit the data.

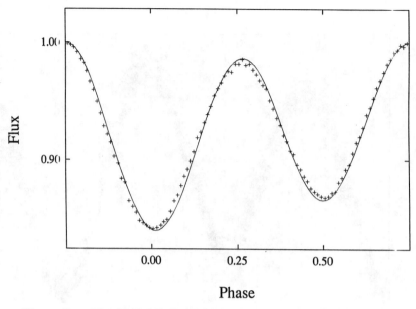

Figure 5. The Model R Band Light Curve of GSC 3493_1097

7. GSC 3493_1158

Serendipitously the star GSC 3493_1158 was included in the field and observed to have both primary and secondary eclipses. From the two primary eclipses and requiring the secondary eclipses to occur at 0.5 phase we find the following ephemeris:

HJD of Minima = 2450954.00(10) + 1.228(6) × E.

The differential R band light curve of GSC 3493_1158 can be seen in Figure 6. The precision is much less for this star because it is sixty times fainter than the comparison star whose peak brightness nearly filled the CCD's well.

Displayed in Figure 7 is an example light curve of this system, made in similar manner as to that used for star GSC 3493_1097. Again the temperature of the larger hotter star was assumed from the (B-V) to be 6300K. The temperature of the smaller cooler star then would be 3030K±200K. This is the temperature of approximately an M3V star which would give a mass ratio of about 0.2 and a ratio of the radii of 0.43. The fractional radius of the larger star was then adjusted to be 0.37 and the inclination was adjusted to 80° ± 2°.

8. Conclusion

Not all fields we have observed are as rich in variable stars as this one was, but that of RE J0725-002 (Robb and Gladders, 1996) was its equal. High precision CCD photometry will help us to discover many new variable stars and possibly to find new kinds of low amplitude long period variable objects.

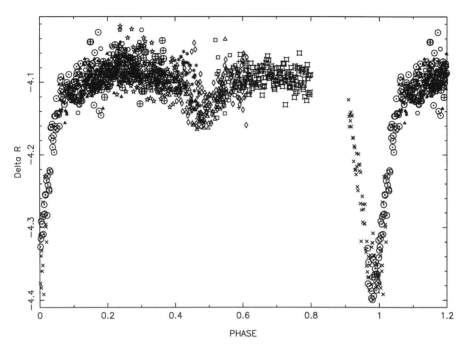

Figure 6. The Differential R Band Light Curve of GSC 3493_1158

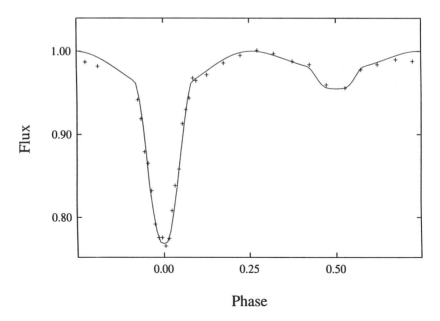

Figure 7. The Model R Band Light Curve of GSC 3493_1158

References

Beers, T.C., Bestman, W., Wilhem, R., 1994, AJ, **108**, 268

Bradstreet, D.H., 1993, Binary Maker 2.0 User Manual, Contact Software, Norristown, PA 19401-5505, USA

ESA, 1997, The Hipparcos and Tycho Catalogues, ESA SP-1200

Jenkner, H., Lasker, B., Sturch, C., McLean, B., Shara, M., Russell, J., 1990, AJ, **99**, 2082

Kwee, K.K., van Woerden, H., 1956, *BAN*, **12**, 327

Robb, R.M. and Gladders, M.D., 1996, IBVS 4412

Robb, R. M. and Honkanen, N. N., 1992, in *A.S.P. Conf. Ser.*, **38**, Automated Telescopes for Photometry and Imaging, ed. Adelman, Dukes and Adelman, 105

Rucinski S.M., 1997, AJ, **113**, 407

Stephenson, C.B., 1986, AJ, **91**, 144

Urban, S.E., Corbin, T.E., and Wycoff, G.L., 1998, AJ, **115**, 2161

Weis, E.W., 1991, AJ, **102**, 1795

Toward Precision Photometry of Red Variable Stars

Therese A. Ostrowski, Robert E. Stencel

University of Denver, Denver Colorado

Abstract. Among the discoveries reported by the Hipparcos team, was the detection of short-term photometric fluctuations in several long period or Mira variables (de Laverny et al. 1998 Astron. Astrophys. 330: 169). Nearly 15 percent of the 250 Miras surveyed in the broadband 380 to 800 nm filter showed variation of 0.2 to 1.1 magnitude on timescales of 2 to 100 hours, preferentially around minimum light phases.

We have begun an observational effort to confirm these variations. We seek to correlate the behavior with ancillary information, such as optical and infrared spectra plus maser data, to determine whether the fluctuations can be understood as thermal instabilities or so-called molecular catastrophe (Muchmore, Nuth and Stencel 1987 ApJ 315:L141; Stencel et al. 1990 ApJ 350:L45). A progress report and call for observation coordination are given.

1. Introduction & Background

Variable stars are stars that change in brightness/magnitude. Depending on the type of variable star, these brightness changes can range from a few hundredths too as much twenty magnitudes over periods of a fraction of a second to years. The American Association of Variable Star Observers (AAVSO) reports that there are now over 28,000 stars known to be variable and 14,000 more that are suspected to be changing in brightness in our own galaxy.

There are many types of variable stars, ranging from pulsating, eruptive, eclipsing variable, to rotating variables. Research on variable stars is important because it can provide much information about stellar properties, such as mass, radius, luminosity, temperature, internal and external structure, composition, and age.

In the category of pulsating variables there are Long Period or Mira-type variables. Mira-type variable stars are very cool red giants with temperatures around 3000 K, they are very large, ranging 200 to 300 times the radius of the sun, and very luminous, 3000 to 4000 times the luminosity of the sun. They are post-main sequence, pulsating long-period variables, with periods ranging from 150 to 1000 days. They are known to have large amplitudes of light variation of more than 2.5 magnitudes in the visual and more than 1 magnitude in the infrared wavelengths.

These are the types of stars of interest for this study. The main goal of this student project is to select a few of these particular stars, and observe them over a period of time using CCD technology. These images will be analyzed using aperture photometry and the results will be compared to know findings. The results of this study will determine if we can match the variability (magnitude fluctuations) of the stars as seen in previous studies.

2. The Hipparcos Variables

The majority of the stars selected for observation came from the Hipparcos satellite mission. Part of this mission was to make position measurements (astrometry) and record brightness measurements (photometry) of all its 118,000 target stars. The satellite was named after the Greek astronomer Hipparchus, who in 129 BC completed a catalogue of a thousand stars. By noting their directions in the sky and their relative brightness, he founded the science of astrometry, or star measurement. The stars in this study include T Hya, X Hya, R Leo, etc. R Leo, although it is not a Hipparcos variable star was CCD imaged and analyzed in order to get a baseline reading when using the software MIRA, for determining magnitudes. R Leo's coordinates: Right Ascension (09 47 33.44), Declination (+11 25 45.8), Spectral Type (M8.0), Period in hours (7501.68), B Magnitude (7.32), V Magnitude (6.02).

Table 1 represents a number of Mira-type variable stars that were used in a study conducted by Laverny, Mennessier, Mignard and Mattei (1998 A & A 330:169). This special class of variable stars were found to have rapid variations in amplitude ranging from 0.23 magnitude up to 1.11 magnitude with timescales extending from 2 hours up to almost 6 days (albeit, with limited sampling). According to Laverny, Mennessier and Mattei, these variables were also found to be oxygen-rich. The authors also theorize that such variations might be related to molecular opacity changes and then to variations in the physical conditions of the stars, inducing instabilities. However, other mechanisms might be invoked, like hydrodynamic effects.

3. Observations & Analysis

Data is being gathered using the "Roboscope" (a semi-automated 8-inch Schmidt-Cassegrain telescope) at the Chamberlin Observatory, Denver, Colorado. An SBIG ST5 CCD camera is used to take the pictures. This camera has a dynamic range of 0 to 2^{14} numerical capacity and it is able to display 16,384 shades of gray (14 bits). SKY PRO software, (Software Bisque), is used in order to obtain the digital pictures taken with the camera. Aperture photometry (measuring the instrumental magnitude of the star) is then performed using the software program MIRA. This method measures the total brightness of one or more objects using an automated background subtraction method. This method involves summing the values of all pixels within a defined boundary (circular) centered on the star, then subtracting the contribution from the diffuse sky background, usually by sampling an annulus of pixels around the star. Data will also be collected from Denver University's new Mt. Evans Observatory (www.du.edu/~rstencel).

4. Results and Discussion of Accuracy

Observations are underway. To this date, multiple pictures have been taken of the Mira star, R Leo and several stars in Table 1. Using the software program MIRA, and its Aperture Photometry routines to evaluate magnitudes differentially from the CCD frames.

Table two is a listing of the CCD images taken of R Leo to date. The date, exposure time of the image, along with the magnitude found for R Leo and two comparison stars are listed. Due to software conversions the comparison star #2 was unobtainable in some of the images. Using the Aperture Photometry procedure in MIRA the instrumental magnitudes were found to be negative, these have not been calibrated yet.

Table three displays the differences found between the same star taken from consecutive images on the same night. This table shows that we are not getting a perfectly constant reading from the images, which can be explained by various factors, camera temperature, software calculations, etc.

Table four compares the magnitude between the three stars within the same exposure. As we can see from the table, the mean difference in magnitude found between R Leo and the comparison star 2 was approximately -4.224 ± 0.043 (s^2). The difference between R Leo and the comparison star 3 was approximately -5.084 ± 0.061 (s^2), and the difference between star 2 and star 3 was 0.864 ± 0.109 (s^2).

5. Conclusion

From the data collected so far we find that it is possible to get relatively stable readings from image to image on the same night using CCD technology to approximately 10% (0.1 mag). When observing R Leo over a time span of a month we find that the two comparison stars were nearly stable while there was some variation in R Leo, which is to be expected. Currently an observational study of the Hipparcos variables is underway.

This effort, to date, has proved to be full of promise and we urge others to take part in the observing of the Hipparcos variable stars. Suggestions and comments would be greatly appreciated, (tao@phoenix.phys.du.edu).

Acknowledgments. Special thanks goes to the William F. Marlar Foundation, for the support of Therese Ostrowski's undergraduate research.

References

Laverny, P., Mennessier, M.O., Mignard, F., & Mattei, J.A. (1998). Detection of short-term variations in Mira-type variable from Hipparcos photometry, A & A 330:169.

Mattei, J.A. (1996). Introducing Mira variables. Presented at the AAVSO 85th Annual Meeting.

Wilson, L.A. (1996). "Theoretical Glue": Understanding the observed properties of Miras with the help of theoretical models. Presented at the AAVSO 85th Annual Meeting.

www.AAVSO.org (website)
www.du.edu/~rstencel (DU observatories website)
www.skypub.com (website)

TABLE 1

SELECTED HIPPARCOS VARIABLES UNDER STUDY

Star Name	Right Ascension	Declination	Spectral Types	Period (Days) LVP Mira	Mag (Hp)	Delta Mag	Timescale (Hours)
SV And	00 04 20.03	+40 06 35.5	M5.0 - M7.0	316.2	11.56	0.65	10.66
R Cet	02 26 02.20	-00 01 42.3	M4.0 - M9.0	166.2	11.27	0.26	4.27
T Eri	03 55 13.8	-24 01 58	M3.0 - M5.0	252.3	11.09	0.31	23.47
V Mon	06 22 42.84	-02 11 48.7	M5.0 -M8.0	340.5	11.26	0.32	4.61
RX Mon	07 29 21	-04 00 17.0	M6.0 -M9.0	345.7	12.20	0.42	15.28
T Hya	08 55 39.75	-09 08 29.2	M3.0 -M9.0	298.7	10.86	0.34	11.01
X Hya	09 35 30.3	-14 41 29	M7.0 -M8.5	301.1	10.58	0.26	28.08
RR Boo	14 47 05.69	+39 19 00.5	M2.0 -M6.0	194.7	11.54	0.25	4.27
RT Boo	15 17 14	+36 00 21.6	M6.0 -M8.0	273.9	11.58	0.34	10.67
S Ser	15 21 39.49	+14 18 52.5	M5.0 -M6.0	371.8	8.35	0.32	93.86
X CrB	15 40 53.4	+36 14 52.8	M5.0 -M7.0	241.2	12.42	0.23	2.48
AH Ser	15 59 20	+19 00 47.7	M2.0 -?	283.5	10.96	0.34	96.00
RU Her	16 10 14.5	+25 04 13	M6.0 -M9.0	484.8	11.67	0.55	25.94
SS Her	16 32 55	+06 00 51.5	M0.0 -M5.0	107.4	11.88	0.69	110.93
XZ Her	18 10 03.87	+18 06 15.2	M0.0 -?	171.7	11.44	0.34	6.40
CE Lyr	18 36 52	+28 00 04.4	-	318.0	11.64	0.44	6.74
HO Lyr	19 20 09	+41 00 40.9	M2.0 -?	100.4	12.52	0.27	2.48
AM Cyg	20 48 58	+31 00 50.9	M6.0 -?	370.6	12.96	0.56	138.67

* The Hipparcos magnitude, denoted by Hp is at the precise epoch of observation. The last two rows Delta Mag and Timescale indicate the change in magnitude, and timescale for the change.

TABLE 2

OBSERVATIONS
(Instrumental Magnitudes)

Object	Date & Time (MDT)	Exposure Time (Sec)	Mag RLEO	Mag Star 2*	Mag Star 3*
RLeo429A	4/29/98, 10:56pm	1.0	-14.564	-10.243	-9.392
RLeo429B	4/29/98, 10:58pm	.75	-14.678	-10.417	-9.395
RLeo429B	4/29/98, 11:05pm	1.0	-14.564	-10.243	-9.257
RLeo510A	5/10/98, 10:34pm	1.0	-14.185	-	-9.432
RLeo510B	5/10/98, 10:36pm	1.5	-14.032	-	-9.408
RLeo513A	5/13/98, 10:03pm	1.0	-14.320	-10.292	-9.545
RLeo513B	5/13/98, 10:05pm	1.0	-14.385	-10.453	-9.243
RLeo513C	5/13/98, 10:06pm	2.0	-14.188	-10.526	-9.560
RLeo519A	5/19/98, 10:21pm	1.0	-14.510	-	-9.558
RLeo519B	5/19/98, 10:23pm	1.0	-14.469	-	-9.325
RLeo519C	5/19/98, 10:25pm	1.5	-14.291	-	-9.395
RLeo519D	5/19/98, 10:27pm	1.5	-14.425	-10.493	-9.342
RLeo519E	5/19/98, 10:29pm	0.75	-14.530	-10.348	-9.495
RLeo526A	5/26/98, 10:00pm	1.0	-14.487	-10.243	-9.392
RLeo526B	5/26/98, 10:03pm	0.75	-14.678	-10.417	-9.395
RLeo526C	5/26/98, 10:04pm	1.0	-14.564	-10.243	-9.392

* Normalized to 1 second exposure.

TABLE 3

DIFFERENTIAL PHOTOMETRY RESULTS

For R Leo

Difference Between Normalized Exposures
(Within the Same Night)

Frames	R Leo	Star 2	Star 3
RLeo429A - B	-0.114	-0.174	-0.003
RLeo429B - C	0	0	0.135
RLeo510A - B	0.153	-	0.024
RLeo513A - B	-0.065	-0.161	0.302
RLeo513B - C	0.197	-0.073	-0.317
RLeo519A - B	0.041	-	0.070
RLeo519B - C	0.178	-	-0.070
RLeo519C - D	-0.134	-	0.053
RLeo519D - E	-0.105	0.145	-0.153
RLeo526A - B	-0.191	-0.174	-0.003
RLeo526B - C	0.114	0.174	0.003

TABLE 4

DIFFERENTIAL PHOTOMETRY RESULTS

For R Leo

Difference Between Normalized Exposures

(Same Night)

Frame	R Leo – Star 2	R Leo – Star 3	Star 2 – Star 3
RLeo429A	-4.321	-5.172	-0.851
RLeo429B	-4.261	-5.283	-1.022
RLeo429C	-4.321	-5.307	-0.986
RLeo510A	-	-4.753	-
RLeo510B	-	-4.624	-
RLeo513A	-4.028	-4.775	-0.747
RLeo513B	-3.932	-5.142	-1.210
RLeo513C	-4.662	-5.628	0.034
RLeo519A	-	-4.952	-
RLeo519B	-	-5.144	-
RLeo519C	-	-4.896	-
RLeo519D	-3.932	-5.083	-1.151
RLeo519E	-4.182	-5.035	-0.853
RLeo526A	-4.244	-5.095	-0.851
RLeo526B	-4.261	-5.283	-1.022
RLeo526C	-4.321	-5.172	-0.851

Mean Difference	-4.224	-5.084	0.864
Standard Deviation	0.208	0.248	0.330
Standard Dev. Squared	.043	.061	.109

Comparative Photometric Reduction Techniques

Eric R. Craine

Western Research Company, Inc., Tucson, AZ 85719/ GNAT, Inc., Tucson, AZ 85719

Michael Snowden

Lanka Astronomical Institute, Kalutar, Sri Lanka, Visiting Scientist, SAAO, Capetown, South Africa; Mt. John University Observatory, New Zealand, and GNAT, Tucson, Arizona

Peter Martinez

SAAO, Capetown, South Africa

Abstract. We are in the process of experimenting with CCD imaging photometry using small telescopes in an effort to understand the limitations of such photometry which are likely to be encountered with automatic imaging telescopes (AITs) in the global network of automatic telescopes (GNAT). To that end we have collected several sets of "typical" CCD images intended for photometric reduction and have begun analyzing these images using several different techniques. We provide a progress report here on the early stages of that effort and note some of the conclusions which have been reached to date.

1. Introduction

We are in the process of experimenting with differential photometry reductions using the prototype 0.5-m GNAT, 0.6-m Mt. John Observatory and 1.0-m South African Astronomical Observatory telescopes. We plan to eventually conduct reductions with IRAF, MIRA and SPS software packages (all discussed elsewhere in this volume). Both aperture photometry and growth curve photometry have been undertaken where feasible; the material presented here were derived from the MIRA package.

Seven data sets have been examined:

1. aperture photometry differential growth curves for the Mt. John data,

2. SAAO aperture photometry data for a centrally located comparison star of moderate brightness,

3. SAAO growth curve data for a bright comparison star and critical manual background subtraction,

4. SAAO growth curve data for a bright comparison star and mean automated background subtraction,

5. SAAO growth curve data for a bright comparison star and critical background subtraction,

6. GNAT aperture photometry, and

7. GNAT growth curve photometry for the same field.

In addition, we present empirical photometric error versus signal/noise functions for some of the data sets.

2. Observations

Several CCD images were made at three different observatories as tabulated below:

GNAT, Tucson, Arizona
telescope: 0.5-m R-C reflector
camera: SouthWest Cryostatics thermoelectrically cooled TK512 CCD
filter: Johnson V
calibration: twilight flats
observations: M67

Mt. John University Observatory, New Zealand
telescope: 0.6-m B&C reflector
camera: thermoelectrically cooled 512x512 CCD
filter: Stromgren V
calibration: dome flats
observations: field in NGC 2516 (stars Cox 38 and Cox 93)
        observations on March 24, 25, April 21, 23 and 24, 1997

South African Astronomical Observatory, Capetown, South Africa
telescope: 1.0-m reflector
camera: liquid nitrogen cooled Tek 8 CCD
filter: Johnson V
calibration: twilight flats
observations: field in NGC 2516 (stars Cox 38 and Cox 93)
        observations on December 28, 30, 1995 and January 10, 12 and 14, 1996

3. Data Reduction

All of the images obtained for this project were formatted as standard FITS images, and were stored either on CD-ROM or on ZIP disks. For the purposes of this report only the MIRA program was used for data reduction. MIRA does not operate on standard FITS format, but rather converts FITS images to a MIRA format (apparently a version of FITS); this is easily accomplished through a translation utility provided with the MIRA package.

The file naming of CCD image data is a subject worthy of protracted discussion in its own right, and did occasion a great deal of discussion during this project. The issue is that CCD photometry reduction requires multi-step processing on the original input images and can quickly mushroom into handling of a very large number of images. For our purposes we wanted to be able to look critically at the data at each step along the way, thus rendering automated or batch processing of the data of more limited value (and for which many of the intermediate images could be discarded, thus obviating the need to be concerned about naming). Numerous conventions are possible, we settled upon preserving the original file name and using a file extension to indicate the operations which had been performed. Thus if the base filename of an image was fn, then we adopted the following: fn.fts = raw original FITS format image, fn.1 = raw MIRA format image, fn.2 = dark subtracted MIRA format image, fn.3 = flat-fielded, dark subtracted MIRA format image, etc. This, of course, sidesteps the issue of what the detailed structure of fn should be, but we will leave that for another venue. We do believe this is an important issue worthy of consideration, but that sentiment anticipates what one of our conclusions will be in the last section of this report.

In general a fairly standard pre-processing procedure was followed for all of the images. Dark frames were averaged to create a master dark, which was then subtracted from each of the flat fields and program images. The program images were then flat fielded by dividing them by the flat field image. These pre-processed frames were then each individually examined for gross abnormalities which would serve as clues to processing errors. As a point of interest we quickly confirmed what has been stressed by many others, i.e. a visually pleasing image is not necessarily very useful, and a rather ugly image is quite capable of containing valuable photometric information. In the case of our images this was most often a function of signal-to-noise ratios in the images, which we soon came to appreciate could be unacceptably low for some otherwise very nice looking images. The lesson to remember here is that just because the image looks good, don't infer that it should produce good photometry.

3.1. Aperture photometry differential growth curves for the New Zealand data

The program field for the NGC 2516 observations is shown in Figure 1.

The Mt. John Observatory data, for five nights, were reduced using the aperture photometry routine. For each night the dark images were averaged by co-adding them and dividing the result by the number of dark images. During the observing run dark images were made at the same integration times as those used for program images, so that scaling as a function of integration time was not required during the data reduction.

Twilight flats were averaged by the same technique, and both the mean flat and the set of program images were then dark subtracted. Bias frames were not taken in this observing program. For all of these operations real arithmetic was used rather than integer. The plan was to produce growth curves by using a series of apertures for each of the program stars, in conjunction with the sky annulus for background subtraction. This process is rather slow since the

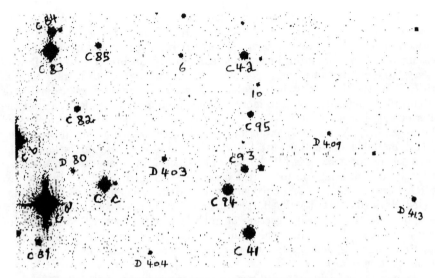

Figure 1. The NGC 2516 program field with identifications of field stars of interest (C for Cox and D for Dachs, single number identifications are anonymous (ANON)).

program will only handle one star aperture at a time, but any number of program stars can be processed simultaneously for that aperture.

Typically about 25 apertures were used, in 0.5μ steps from $2\text{-}14\mu$ radius. A table of instrumental differential magnitude was produced for each star for each aperture size (differential in the sense that the zero point of the magnitude scale was chosen as an arbitrary value). This magnitude was then plotted as a function of the star aperture, thus yielding a growth curve for the star of interest. Figure 2 presents these growth curves for individual stars, in order of apparent brightness (Figures 2a-2e).

In Figure 2f we show differential photometry of two check stars in the field with respect to a star located in the center of the field.

3.2. SAAO aperture photometry: centrally located comparison star of moderate brightness

SAAO images of NGC 2516 were obtained on the nights indicated in Observations above; twilight flats were obtained, but dark and bias frames were not. The process of data reduction was initially to use aperture photometry, exactly as described above for the Mt. John Observatory data, with the exception that there was necessarily no dark subtraction. The curves of growth are plotted in Figure 3 in the same way as for Figure 2 above. For differential magnitudes with respect to Cox 95, absolute values of differential magnitudes are used; the stars are listed from brightest, Cox 41, to faintest, Anon 10.

Comparative Photometric Reduction Techniques   217

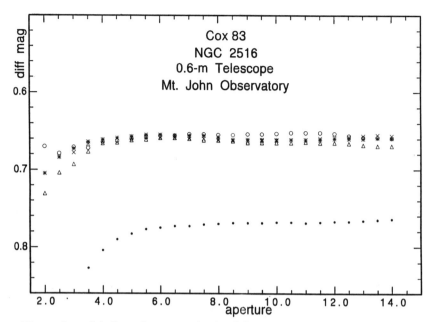

Figure 2. (a) Growth curves obtained from aperture photometry for Cox 83 over a period of five nights. · = March 24, ○ = March 25, ⋆ = April 21, x = April 23 and △ = April 24, 1997; aperture size in microns.

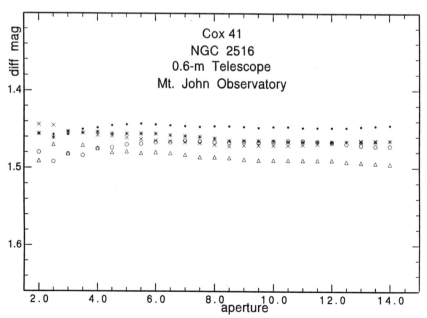

Figure 2. (b) Same as Figure 2a, but for Cox 41.

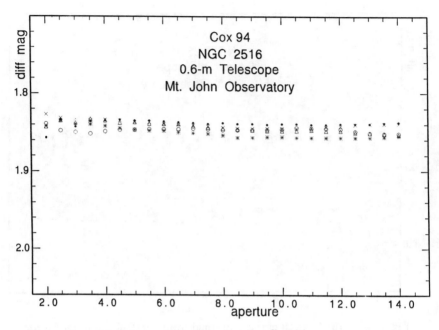

Figure 2.   (c) Same as Figure 2a, but for Cox 94.

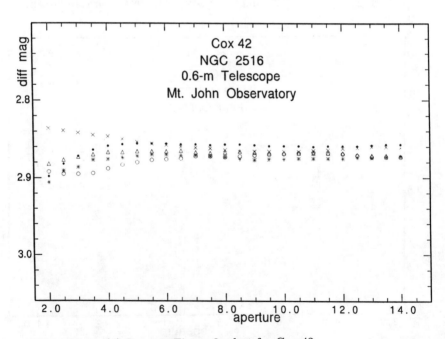

Figure 2.   (d) Same as Figure 2a, but for Cox 42.

Comparative Photometric Reduction Techniques 219

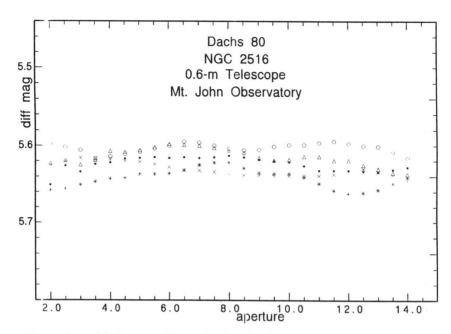

Figure 2. (e) Same as Figure 2a, but for Dachs 80.

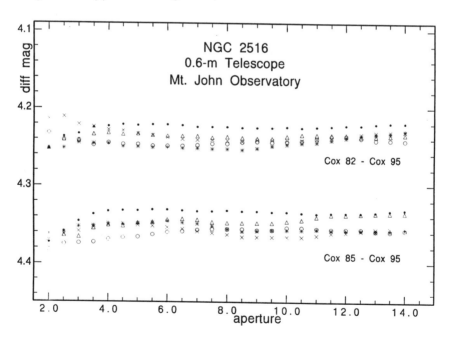

Figure 2. (f) Differential magnitudes with respect to comparison star Cox 95, as a function of aperture size in microns.

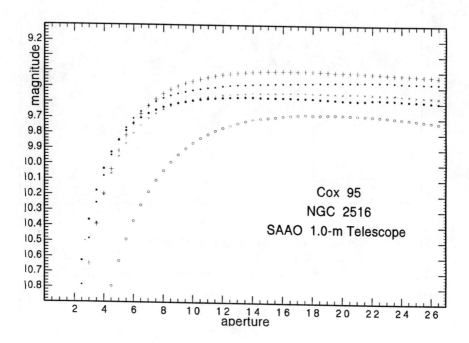

Figure 3. (a) Growth curves obtained from aperture photometry for Cox 95 over a period of five nights. • = 12/28/1995, · = 12/30/1995, + = 1/10/1996, x = 1/12/1996 and ○ = 1/14/1996; aperture size in microns.

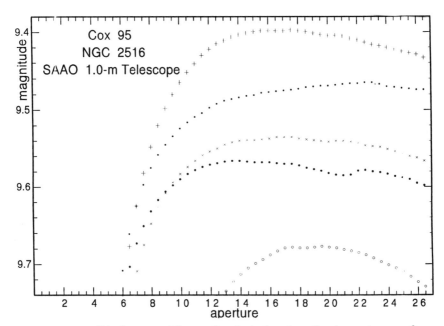

Figure 3. (b) Same as Figure 3a, but showing the importance of using a more restricted scale.

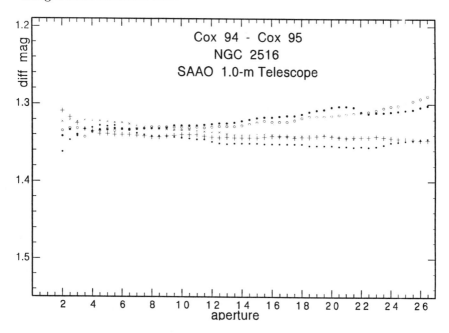

Figure 3. (c) Differential magnitudes with respect to comparison star Cox 95; for Cox 94.

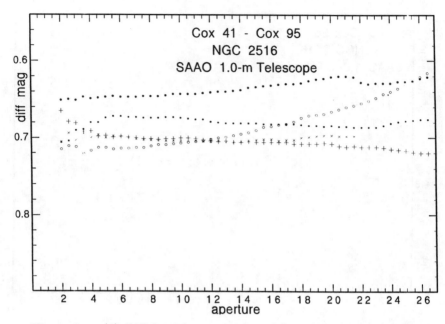

Figure 3. (d) Differential magnitudes with respect to comparison star Cox 95; for Cox 41.

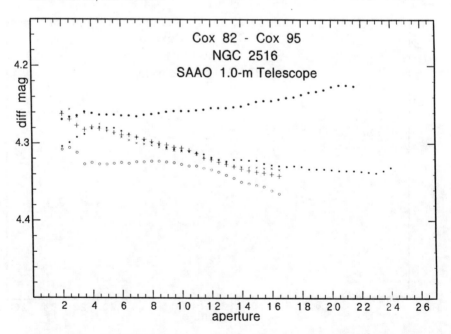

Figure 3. (e) Differential magnitudes with respect to comparison star Cox 95; for Cox 82.

Comparative Photometric Reduction Techniques 223

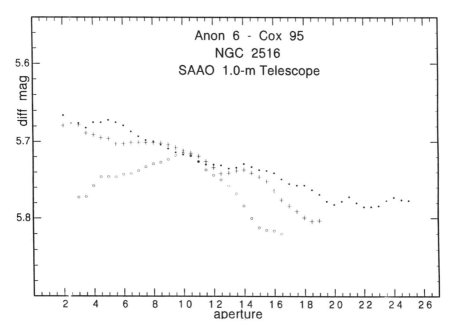

Figure 3. (f) Differential magnitudes with respect to comparison star Cox 95; for Anon 6.

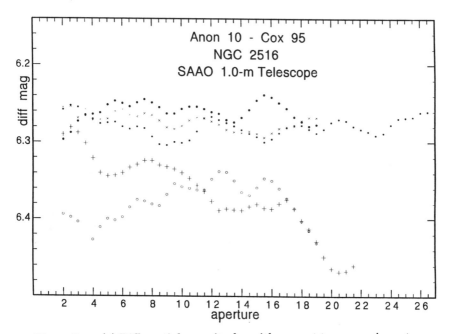

Figure 3. (g) Differential magnitudes with respect to comparison star Cox 95; for Anon 10.

3.3. SAAO growth curve data

The SAAO images were subsequently reduced using the growth curve routine in the MIRA package. The procedure was to flat-field the individual frames using a mean flat. The mean flat was obtained using the combine images routine. The mean flat for the SAAO data has a very high reading and must be drastically reduced to a background level near 1.0. This is accomplished by scanning the mean flat to find the background level which is slightly less that the minimum level displayed in the image. For a typical frame it was about 33,000. Using the MIRA arithmetic routine the frame was then divided by 33,000 and saved under the same name. This process of creating mean flat fields was repeated for all nights.

Growth curve photometry was then implemented for each star in the frame, and, using an iterative process of selecting background values to subtract, the growth curve was examined until it achieved the best possible plateau. In Figure 4 we show the results of this approach.

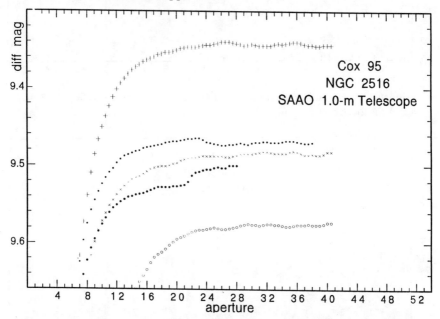

Figure 4. (a) Growth curves obtained from growth curve photometry for Cox 95 over a period of five nights. • = 12/28/1995, · = 12/30/1995, + = 1/10/1996, x = 1/12/1996 and o = 1/14/1996; aperture size in microns.

Comparative Photometric Reduction Techniques 225

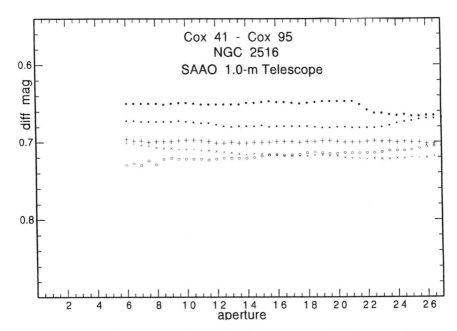

Figure 4. (b) Differential magnitudes with respect to comparison star Cox 95; for Cox 41; compare with Figure 3d.

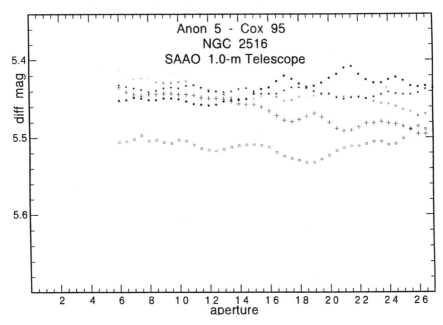

Figure 4. (c) Differential magnitudes with respect to comparison star Cox 95; for Anon 5.

3.4. SAAO growth curve data: brighter comparison star; mean background subtraction

The same data reduction described in 3.3 above was done using a brighter star, Cox 94, as the primary comparison for differential photometry. In this case mean values of the background were used for the entire image. Sample results are shown in Figure 5.

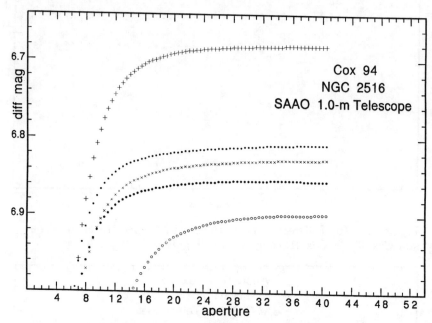

Figure 5. (a) Growth curves obtained from growth curve photometry for Cox 94 over a period of five nights. • = 12/28/1995, · = 12/30/1995, + = 1/10/1996, x = 1/12/1996 and o = 1/14/1996; aperture size in microns.

3.5. SAAO growth curve data: brighter comparison star; critical background subtraction

Finally, the SAAO data were again reduced, exactly as in 3.3 above, except that the brighter comparison star, Cox 94 was used for the differential photometry. Results are shown in Figure 6.

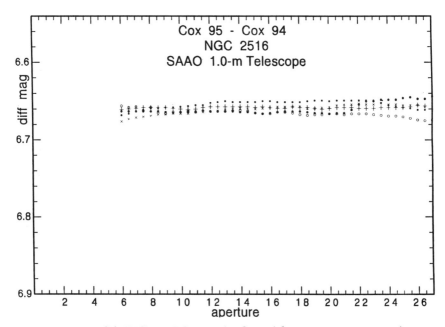

Figure 5. (b) Differential magnitudes with respect to comparison star Cox 94; for Cox 95.

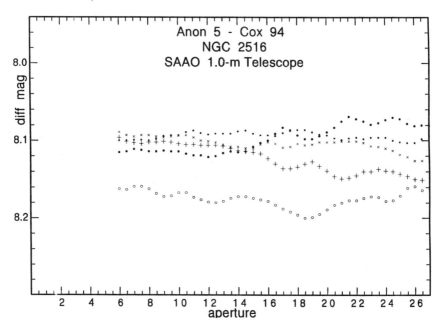

Figure 5. (c) Differential magnitudes with respect to comparison star Cox 94; for Anon 5; compare with Figure 4c.

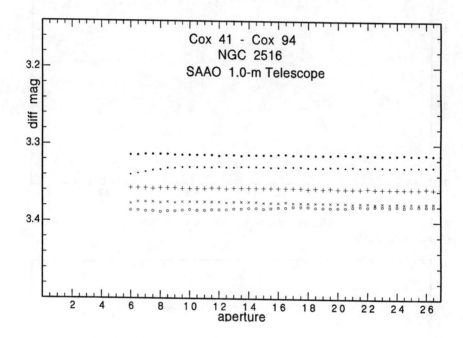

Figure 6. (a) Differential growth curves obtained from growth curve photometry for Cox 41 with respect to Cox 94 over a period of five nights. • = 12/28/1995, · = 12/30/1995, + = 1/10/1996, x = 1/12/1996 and ○= 1/14/1996; aperture size in microns. Compare with Figure 4b.

Comparative Photometric Reduction Techniques 229

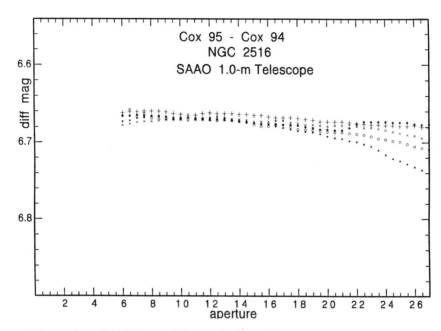

Figure 6. (b) Differential magnitudes with respect to comparison star Cox 94; for Cox 95.

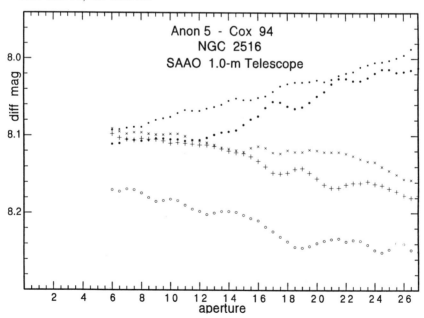

Figure 6. (c) Differential magnitudes with respect to comparison star Cox 94; for Anon 5; compare with Figure 4c and 5c.

3.6. GNAT aperture photometry

Observations of a 15x15-arcmin field in M67 by the GNAT 0.5-m automated telescope were reduced using the same techniques as described in 1. above for the Mt. John Observatory data. Of significant difference was the much greater field crowding of the M67 image, in part due to the greater number of stars in that field, and in part due to differences in the image plane scale of the two telescope-camera systems.

Several field stars were measured and differential magnitudes relative to one star, Racine A, were graphed as a function of stellar aperture size. Figure 7 presents these growth curves for individual stars, in order of apparent brightness (Figures 7a-7d).

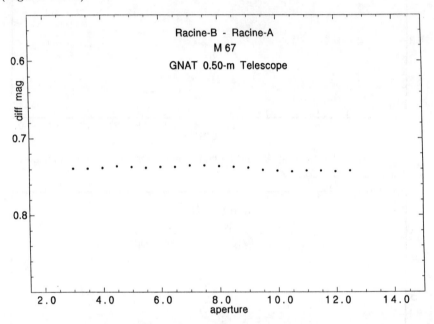

Figure 7.   (a) Growth curve for Racine B.

3.7. GNAT growth curve photometry for the same field

The same M67 GNAT field described in 3.5 above was re-measured using the curve of growth photometry routine. In this instance, one star in the field is measured at a time, with calculations being made for all of the stellar apertures at once. The background subtraction is accomplished by the operator manually selecting a value to subtract, generally guided by obtaining statistics on the sky background in one or more selected areas surrounding the star. This process can be iteratively undertaken, until such time as the resultant curve of growth appears to present the most stable result. The process is then repeated for each star in the field.

This technique is slow and labor intensive, but provides a remarkably sensitive probe of the local background behavior of the image and, particularly for

Comparative Photometric Reduction Techniques 231

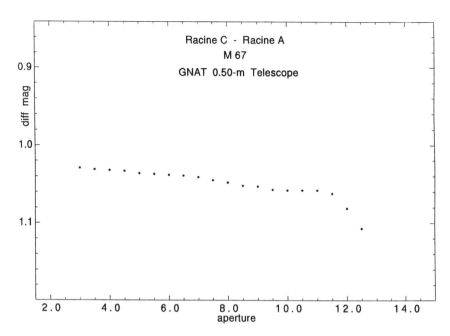

Figure 7. (b) Growth curve for Racine C.

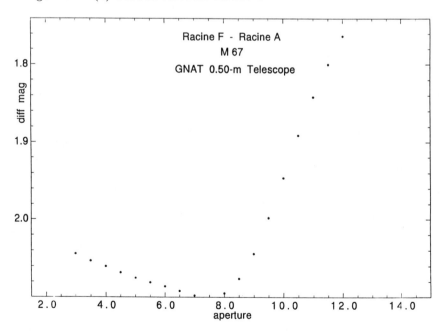

Figure 7. (c) Growth curve for Racine F.

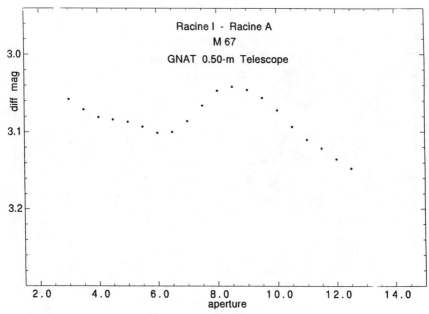

Figure 7.   (d) Growth curve for Racine I.

crowded fields, can be far superior to simple aperture photometry using a sky background annulus.

The same stars as in Figure 7 are presented in Figure 8 for purposes of comparison and discussion. Because this is a fairly crowded field many of the growth curves rise to a plateau for the program star, remain on the plateau as the apertures become larger, then rise again a one or more additional stars are encompassed by the apertures. With the growth curve photometry routine the user manipulates the value of the background selected until a flat plateau is achieved for the program star. It can be seen by comparing Figures 7 and 8 that for the aperture photometry the plateau is never properly achieved, but rather the growth curve turns over, or continues to rise. This is symptomatic of incorrectly calculated sky background, which can have a significant impact on the resultant instrumental magnitude.

Finally, for the GNAT data we have compared the photometric errors (standard errors about the mean over selected ranges in the growth curves) with the signal-to-noise ratio in specific stellar images. These data are shown in Figure 9 below.

4. Summary

To evaluate the photometric precision attainable with a system it is not uncommon for people to look at a theoretical relationship relating photometric error to signal-to-noise and assume that this represents their typical errors. Indeed, there are numerous papers in the literature where exactly this has happened. In

Comparative Photometric Reduction Techniques 233

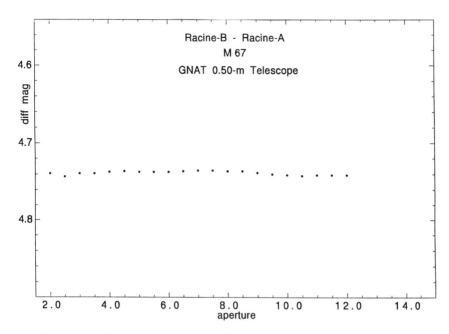

Figure 8.   (a) Growth curve for Racine B.

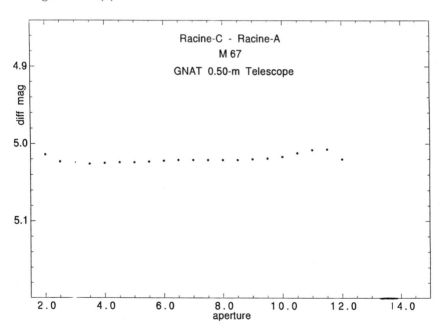

Figure 8.   (b) Growth curve for Racine C.

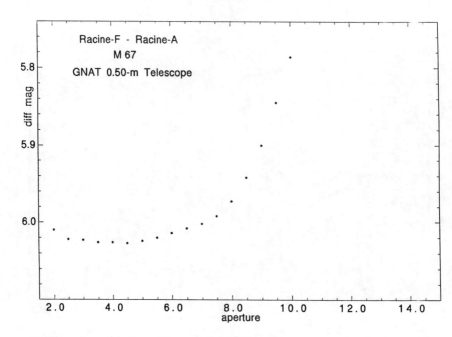

Figure 8.   (c) Growth curve for Racine F.

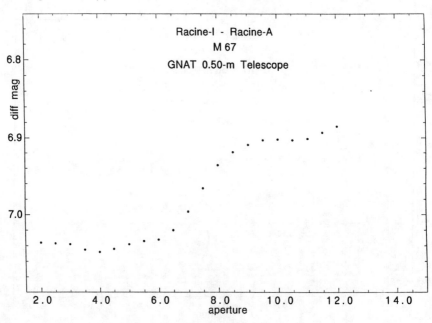

Figure 8.   (d) Growth curve for Racine I.

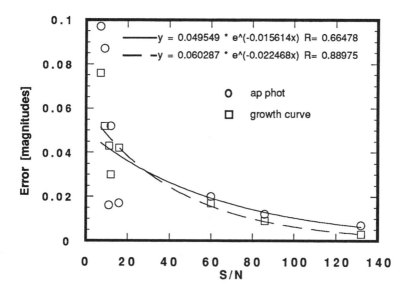

Figure 9. Empirical photometric error as a function of signal-to-noise ratio.

fact, one is well advised to experiment with several techniques for reducing data and perform enough reductions to be able to statistically evaluate the errors.

In these early stages of reducing small telescope CCD photometry we are attempting to illustrate many of the issues which confront a first-time aspirant to high precision photometry. We hope that newcomers to CCD photometry will find study of the data presented her helpful in answering some of the questions which inevitably arise. Some of what can be learned from these data are presented below, in no special order.

1. When this workshop was convened several expressed the opinion that the acquisition of high precision photometry was a well-understood process (although others also argued that high precision CCD photometry is an oxymoron !). During one of the longest and most divisive discussion sessions the pros and cons of obtaining biases, darks and flats were argued. One school has emerged which contends that these corrections are not needed for precision differential photometry. Our experience has been to the contrary and we contend that to attain high precision photometry of any type it is essential to obtain large numbers of biases, darks and flats. Certainly, without such calibration milli-magnitude photometry which is repeatable and believable does not seem possible, and we encountered instances in which neglect of these calibrations can result in errors of hundredths of a magnitude, even among high signal-to-noise objects.

2. There are numerous approaches to reducing photometric data, and certainly there has been considerable discussion of the advantages and dis-

advantages of each. Our experience has been that for those just starting out in learning these methodologies, the growth curve photometry routine is by far the most sensitive and can consistently provide more insight into what is actually going on with your image. You may elect ultimately another approach, but if you are truly concerned about the quality of your data, and resulting photometry, the time spent to investigate the growth curves in detail is well spent.

3. The inescapable fact is that quality photometric data reduction is boring, tedious and labor intensive. There is an enormous and understandable desire to automate the process and various attempts have been made in that direction. In reality, automated reduction can be the salvation of a large photometry program, however, to start with automated reduction is, in our opinion, a dangerous error. To understand the nature and quality of the data it is essential that the photometry reduction at least begin with extensive manual interaction with not only each image, but each star in the image. To believe that you can trust and understand your results, in the context of high precision photometry by doing anything otherwise is self-delusional; by way of example, see 4 below.

4. In talking with several astronomers since the initiation of this project, we have been surprised at the rather large percentage of them who do photometry by loading their images into a utility program (MIRA, APPHOT, etc.), select an aperture which someone has told them to use, or which seems about right to them, or whatever, running the program and producing a table of instrumental magnitudes. They then form differential magnitudes with respect to a star, or group of stars, in their image and attach errors which are typically the square root of the signal-to-noise ratio. Examination of almost any of the growth curves in the Figures presented in this paper immediately illustrate the problem; even changing aperture by a small amount can often easily modify the resultant differentials by certainly several thousandths, if not hundredths, of a magnitude. Selecting a single aperture of all of the stars in the field is a tricky business, and batch processing with such an aperture frequently allows for no means of determining the validity of the result.

5. When doing differential photometry in a sparsely populated field, and using a single comparison star (or couple of comparison stars), it is advantageous to choose a high signal-to-noise ratio comparison star (compare, for example, the data sets for Cox 41 above).

6. When doing differential photometry in a moderately densely populated field it is often very easy, and much superior, to make the comparison with an ensemble mean.

7. Growth curve data provides an interesting and instructive way to follow the behavior of variable stars. For example, in the data sets presented here Cox 41 is a variable star whose brightness variations can be seen very well in the night-to-night growth curves.

8. It is very important to keep in mind that photometric precision is a function of signal- to-noise in the star image. We have noticed that there is a particular tendency among those astronomers who learned astronomical imagery with photographic emulsions to judge an image on its appearance. Specifically, if a CCD image looks nice then it should be easy to do good photometric measurements. In fact it comes as a bit of a surprise to see how nice and sharp a low signal-to-noise star image can look, and still be useless for high precision photometry. This is particularly important to remember if you are aspiring to make high precision measurements of a large number of stars (of varying brightnesses) in a single image.

5. Acknowledgements

The authors would like to acknowledge the contributions of the several observatories at which these various observations have been made. We also thank the many colleagues who have shared their (often conflicting!) ideas on how to reduce CCD data.

Differential Photometry Using the GNAT 0.5 m Prototype

J. M. Taylor

Harvard University, Cambridge, MA 02138 USA

E. R. Craine

GNAT, Tucson, AZ 85719 USA

M. S. Giampapa

NSO, Tucson, AZ 85719 USA

1. Introduction

Planet searches in general are both observationally intensive and precision limited. The radial velocity method has been quite successful in such extra-solar planet detections; however, with the maturation of CCD technology and an automated network of telescopes, another method might be pursued. If the plane of ecliptic is edge on, and a planet is in the system, it will appear to transit across the surface of the star; this causes a drop in the received light. If the conditions are right (the cross section of the planet large enough compared to the star to create a photometrically noticeable drop in light) and the star well monitored both before and during the event, such a transit event could be detected.

Giampapa, Craine, and Hott (1995) investigated the probability of success for such a search. For maximum detectability, all the program stars will be M dwarfs with Jupiter size companions. In such a system, the radius of the planet is approximately a tenth of the star's radius, and thus the percentage light drop, as given by

$$\Delta L = \left(\frac{R_{\text{planet}}}{R_{\text{star}}}\right)^2, \quad (1)$$

would be 1%. If such a level of photometric accuracy could be achieved in multiple filters, and if a system existed that could handle observing enough such targets, a search might be successful.

With the Global Network of Automated Telescopes (GNAT), such an observationally intensive program could be carried out. However, whether it could achieve the level of accuracy necessary (say 1% at five sigma) for enough targets, observed often enough, remains to be seen. Young, et al. (1991) investigated the many problems with millimagnitude photometry using photomultiplier tubes. Many of the atmospheric, telescope, and filter considerations are valid for CCD observations as well. In particular, they noted the strong effects of color terms in the extinction, and limitations placed upon accuracy by a filter set (such as Johnson UBV) that does not adequately determine the energy distribution of two stars; differences as high as 2 millimagnitudes could be seen over a night between two stars with the same color index but different energy distributions.

This effect is stronger in some cases than the scintillation noise, which would otherwise be the limiting factor in an observation. (For a discussion of scintillation, see Young (1974)).

Gilliland and Brown (1988), using the then No. 2 0.9m telescope at Kitt Peak, observed the old open cluster M67 looking for stellar variability. By spending a significant amount of time understanding their detector and correcting for its deficiencies, as well as using differential photometry with empirical corrections for extinction, reddening and any other atmospheric effects, they found that over a night of poor conditions (high cirrus clouds with a bright moon) they could achieve a time series sigma of 2.3 millimagnitudes.

In a later survey using five 1m class telescopes, they could achieve an accuracy of 0.8 millimagnitudes per observation within a night under the best conditions, limited only by the shot noise, sky levels, scintillation, and CCD problems (Gilliland, et. al. 1991). However, they were not interested in the night to night stability of the data, a very important factor in any intensive, multi-field planetary transit search program such as the one described in Giampapa, Craine, and Hott.

With the goal of achieving night to night observations only limited by the statistical noise (the error in the photometry), several observations of the old open cluster M67 were taken using the prototype 0.5m GNAT telescope, and were reduced and examined. Both PSF photometry using Stetson's DAOPHOT II and aperture photometry were pursued, to attempt to find the most stable solution. Ideally, the millimagnitude level reductions necessary for a planetary search could be carried out as the data are taken; then, if a transit event is detected, the telescope could stay on the object, and, using the global network, pass off observations before the object sets. Such an automated approach is likely to work better with the 'dumb' aperture photometry, which requires fewer parameters to set and is more forgiving of changes in seeing and variability of the PSF. Unfortunately, the maximum number of targets is much smaller when using aperture photometry compared to PSF photometry.

Finally, after reducing the data, variation is looked for in the sample of stars, and night to night light curves are considered.

2. Observations

We observed several test frames of M67 on the nights of April 12, 13, 17, 19 and 20, 1998. The SITE 512 X 512 CCD was used, cooled by a thermoelectric device with additional cooling provided by running water (as described in this volume in the poster-paper of P. Craine). The detector covered a 15' x 15' field of view, with 1.76" pixels. A total of 15 usable frames (those that were in focus and with high signal to noise) nearly centered on the dipper asterism were available for study from that data. There were six frames with two minute exposures and nine frames with four minute exposures. All observations were taken through a Johnson V filter.

3. Reductions

Keeping in mind the ideals of precision photometry, the CCD had to be calibrated as well as possible.

Several bias frames were analyzed, and a slight structure along the top rows of the chip were noted. However, this structure amounted to, on average, variations of one count. This slight structure was not subtracted from the final images to maintain a high signal to noise. However, the row to row variations in the overscan were quite significant (5 counts) and were fitted with a fifth order polynomial and subtracted off.

Due to the TE cooler used on the chip, average camera temperatures were high compared to liquid nitrogen cooled systems, and the dark counts were approximately 30 counts per minute (or 200! e^- per minute, 5000 times worse than the T2KA at Kitt Peak). However, for an automated system with no observer on site, a system that requires less maintenance is preferable, and the TE cooling system fulfills those conditions. Some overall structure in the darks was noted. The level at the bottom of the chip was uniform, with only the counting, read noise, and digitization errors contributing to the overall noise. The histogram (figure 1) shows a nice gaussian distribution of counts peaked around 30 counts per minute (or 60 counts in this two minute exposure). However, as one progressed towards the top, this split into two gaussians, one centered at 60 counts per minute, one at 30 counts per minute (figure 2). A master dark was compiled out of 18 frames, and used to subtract the overall structure and dark current. This dark had a statistical error of less than one count, thus adding at most one count of error to the reduced frame. This can be accounted for in the analysis using predefined IRAF tasks as an increased readnoise, given by

$$R_{\text{eff}}^2 = R_{\text{detector}}^2 + \sigma_{\text{dark}}^2 \qquad (2)$$

As there is currently no acceptable dome flat system for the GNAT prototype telescopes, and a flat accurate over the whole field was desired, twilight flats were used. The night to night stability of the flats was excellent; division of a master flat from one night by that of another revealed an almost perfect Gaussian. The variation is at most 0.5 % (figure 3); with this in mind, and the fact that twilight flats exist for the V filter for only two nights, those two nights were combined into one master twilight flat, used for all of the nights, with a resulting signal to noise greater than 600. However, the 0.5 % true error in the combined frame corresponds to a S/N of only 200, and that was used in the flat field error analysis for the evenings without their own twilight flats.

So, the total S/N of one pixel with n corrected counts, on this CCD with gain g, is given by

$$(S/N)^2 = \frac{1}{(n/g + R_{\text{eff}}^2 + (g^2 - 1)/12)/n^2 + (1/200)^2} \qquad (3)$$

and we take this to be the limiting factor upon the differential photometry. Ideally, we will be able to approach this statistical limit on the level of night to night variations.

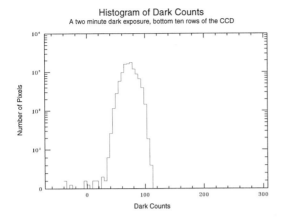

Figure 1. Histogram of the bottom 10 rows of the CCD, after overscan subtraction, for a two minute dark exposure. Note the single gaussian, centered at 60 counts.

4. Analysis

After the frames were cleaned up, they had to be matched up and the intensities for each star determined for each frame. An IRAF task was developed to match the frames by finding the brightest stars in a field, and matching them using a similar triangles matching algorithm to a master list of bright stars. This provided a linear transformation, which was then applied to a master list of all stars in the field. The resulting coordinate transformed list was sent to the various photometry tasks, which recentered within one pixel to allow for non-linear distortions in the CCD.

4.1. Photometry

Both aperture photometry using the IRAF task APPHOT and PSF photometry using DAOPHOT II were used. The FWHM of most stars was 3 pixels (5″), and the field was essentially a crowded one. We wanted to investigate the differences in aperture and PSF photometry for doing this type of differential photometry. For an automated system, with variable seeing and changing conditions, the 'dumb' aperture photometry was likely to work under more conditions than the more sophisticated PSF photometry. However, the crowding of the field might cause problems of varying light leakage from nearby stars as conditions change.

Aperture photometry was pursued using apertures of three, five, and seven pixels. The sky was fit with an annulus with an inner radius of 10 pixels and an outer radius of 20 pixels, which a gaussian was fit to after removing extraneous points (eg stars) and the peak of the gaussian determined the sky value. As the aperture increases, the signal to noise goes up to a certain point. The difference between 5 pixels and 7 pixels was negligible.

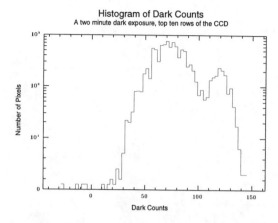

Figure 2. Histogram of the top 10 rows of the CCD, after overscan subtraction, for a two minute dark exposure. Here there are two distributions, one centered at 60 counts and the other at 120 counts.

Analysis of the PSF revealed a quadratic variability across the field, and even using this correction, DAOPHOT II had a 2% residual in the peak. Though the PSF data dealt with neighboring stars quite well, the bad fit created a larger error than the statistical error indicated, and occasionally, a star would 'migrate' to the position of a nearby star, or merge with a nearby star. These mis-matching events added more noise to the data set than the aperture overlaps due to nearby neighbors. Fortunately they were easier to detect than the overlaps, and thus could be taken out of the data.

4.2. Differential Photometry

After performing the photometry, fluxes and 'photometry sigma', that is, the sigma due to the theoretical consideration of all noise sources, were recorded along with the heliocentric Julidan date (HJD). These intensities then needed to be compared to some frame mean, and this frame mean would allow for comparison from frame to frame.

In typical differential photometry, a magnitude is determined for the program star and for several comparison stars that are in every frame. Then, an ensemble magnitude is determined, and this sets the zeropoint of the differential scale.

However, there are several high order extinction effects this doesn't take into account. In particular, Young *et al.* (1991) note that the extinction of any star is given by

$$\Delta Mag = AM - WRMC - W(RM)^2/2, \qquad (4)$$

where A is the extinction coefficient, M is the airmass, W is the bandpass of the filter, C is the color index, and R is the reddening coefficient. The advantage of doing differential photometry falls in the elimination of the first term. Essentially, across a 15' field of view, the worst possible difference in

Differential Photometry Using the GNAT 0.5 m Prototype

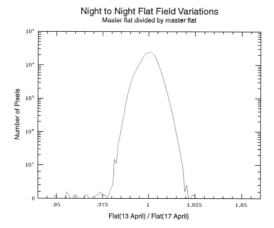

Figure 3. Histogram of twilight flats from one evening divided by those from another.

airmass is approximately 0.005 airmasses (if looking at 45 degrees off zenith or better). Only under very bad conditions does this term contribute at the millimagnitude level.

On the other hand, the color term depends on the difference in color and the *total* amount of airmass being looked through; as a result, it causes a significant change in the differential magnitude of stars of different colors, especially through such broadband filters as Johnson V. Young et al. noted that this effect could be as strong as 1% (0.01 mag).

There are two straightforward ways to combat these effects (the second order extinction and color terms). One could intersperse standard star measurements in between field measurements, and thus determine the A and R coefficients as they changed throughout the night. However, this requires a night stable on the time scale of the spaces between standard star observations. Alternatively, one could adopt the technique of Gilliland and Brown (1988), and empirically correct for spatial and color related extinction variations.

The technique is straightforward. A set of comparison stars is chosen in the field, with the optimal conditions: (1) they need a high signal to noise, (2) they should be well spread over x, y, and color, and (3) they should not be variable stars or have much low level variability. For each star i, an average of that star's flux over all the frames is determined. Then, using the whole set of comparison stars, the relationship

$$I_{i,j}/I_{i,avg} = a_{0,j} + a_{1,j}x + a_{2,j}y + a_{3,j}(B-V) \tag{5}$$

is fit for each frame j using a least squares method. Once this has been done, the corrected intensity for any star i in frame j is given by

$$I_{i,j,corr} = \frac{I_{i,j}}{a_{0,j} + a_{1,j}x + a_{2,j}y + a_{3,j}(B-V)}. \tag{6}$$

This technique is only effective with enough stars to accurately determine the coefficients; for very sparse fields, standard star measurements must be used.

The increased stability of this technique is readily apparent. In figure 4, a histogram of the 'time series sigma' over the 'photometry sigma' is shown. Here, the time series is corrected with standard differential photometry. The ideal case of 'time series sigma' equal to the 'photometry sigma' is hardly realized. However, applying Gilliland and Brown's empirical technique, we get a narrower histogram (figure 5), much closer to being centered at one. The $B-V$ photometry values are taken from Montgomery, Marschall, and Janes (1993) and are not available for all stars; only those stars with $B-V$ values are included in these histograms.

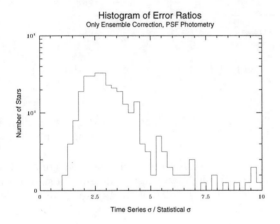

Figure 4. Histogram of the ratio of time series sigma to the photometrically determined sigma, using standard ensemble differential photometry.

To apply this correction to any given field, some setup time is required; in particular, a photometric night must be found to do absolute photometry at the 0.01 magnitude level to get a good color index. And, as noted earlier, there are still fluctuations at around 1 millimagnitude due to the undersampling of the star's energy distribution. This is above the scintillation limit for a 1m class telescope as determined by Young (1974).

For this data, comparison stars were chosen from the set of stars in all frames, such that (1) after a rough fit, they had low variability (their time series sigma similar to their photometric sigma) and (2) they had the highest possible signal to noise. The number of comparison stars used was important, to a point. Figure 6 shows a histogram of the ratio of time series sigma for each star, with 14 comparison stars and with 30 comparison stars. The histogram is wide on the right side, indicating more stars with a greater time series sigma for the 14 star fit than the 30 star fit. However, the same histogram for 30 stars versus 129 stars (figure 7) shows an even Gaussian, suggesting that the differences between the fits is negligible.

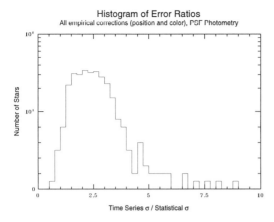

Figure 5. Histogram of the ratio of time series sigma to the photometrically determined sigma, using the empirical fitting to allow for first order position and color effects.

We are fitting for four parameters total, and each should be overdetermined if possible, which means at least three, preferably more data points for each parameter. All of the parameters rely upon good signal to noise data, and each requires a good distribution across the dependent parameter (x, y, and $B - V$). Though for this data set 30 comparison stars do as well as 129, whenever a field is being set up for this technique of differential photometry, several different sets of comparison stars should be chosen and compared; ideally, the best set minimizes the spread and center of the peak of the time series sigma versus photometry sigma histogram. Too many stars will add noise as the signal to noise of the ensemble goes down beyond a certain point, while too few underdetermine the parameters.

After finding a suitable set of comparison stars and performing the empirical fitting, the resulting time series were analysized for variations; the light curves for stars with a high ratio between their time series sigma and their photometry sigma were looked at.

5. Results

5.1. Precision

As a statistical ensemble, the 355 stars in the field allow for analysis of various techniques. The goal is quite simple: a time series sigma equal to the photometry sigma, where only the limitations of the CCD, shot noise, and scintillation limit the signal to noise possible.

The effects of correcting for second order extinction and color extinction on the ensemble have already been shown, and it obviously improves the statistics. The next variable to change is the type of photometry performed. For this situation, PSF photometry should provide the most consistent result, given the

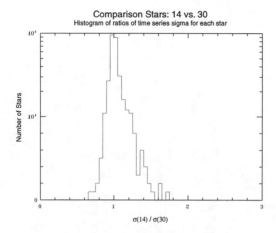

Figure 6. Comparison of time series sigma for a fit determined by 14 stars versus that determined by 30 stars.

crowded nature of the field. However, as noted by Gilliland and Brown, and as we also noticed, for good seeing (3 pixels) the PSF is almost subsampled, and DAOPHOT leaves a 2% residual at the peak. Some sort of PSF subsampling technique would have to be applied to gain subtractions of 0.5% or better; one such technique would be that of Gilliland and Brown.

Looking at a histogram of the time series sigma over the photometry sigma for the PSF photometry, we see that the peak lies at 2.1; thus, the error in the photometry has been underestimated by a factor of 2! A histogram of signal to time series noise reveals no stars with a stability of even 5 millimagnitudes. We need to look for a better technique.

Aperture photometry with a three pixel aperture comes close to ignoring the crowding problem; at that small size, the contribution of nearby stars can be close to negligible. However, analysis of the time series sigma over photometry sigma histogram reveals a histogram with its peak at 1.6, but a wide right side. This is significantly closer to the desired limiting case of one. Looking at its signal to time series sigma (from now on referred to as the signal to noise), the majority of the stars fall with a signal to noise of less than 50; we want at least 300 to have a hope of detecting planets.

The additional error for the 3 pixel aperture might come from a subsampling of the star's flux due to the small aperture size; once again, a subpixel technique might yield better results. Also, the signal to noise issue is in part due to the small number of pixels contributing to the overall flux. The dynamic range of the CCD is already taxed by the field, with several stars beyond saturation, so longer integration times won't help. However, with a larger aperture, a higher signal to noise was achieved. For a 5 pixel aperture, we see two effects.

First, the histogram of time series sigma to photometry sigma is even tighter than the 3 pixel one, though the peak is still around 1.6; we think this is due to the night to night variations in seeing and perhaps from some other sources.

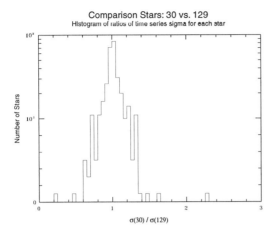

Figure 7. Comparison of time series sigma for a fit determined by 30 stars versus that determined by 129 stars. Note the narrower fit than that of figure 6.

However, without a statistically significant number of frames from any one night, the limits within that night cannot be fully explored. The work of Gilliland and Brown suggest that a ratio of unity could be achieved.

Second, the histogram of signal to noise shows five stars with an accuracy of 0.5% or better. Similarly, looking at the photometry with a seven pixel aperture, there are nine stars with an accuracy of 0.5% or better, and four with 0.3 % or better.

5.2. Variability

Out of this ensemble of stars, those with a high signal to photometry sigma were analyzed for possible variation. Any star with a time series sigma a factor of four or greater than their photometry sigma (constituting a four sigma detection) was flagged as a possible variable star and it's light curve and position on the field examined.

Several variable candidates had nearby neighbors, which caused a 'variation'; for the aperture photometry, changes in conditions led to differing contributions from nearby neighbors. As for the PSF photometry, occasionally a star would 'migrate' to a nearby companion star, drastically changing the light levels. Both types of stars were discarded as non-variables. An example of star suffering from both effects is given in figure 8. The PSF photometry is much more accurate, with only two data points widely divergent from the average value. The aperture photometry, on the other hand, changes dramatically (it's flux changing by a factor of two, equivalent to a magnitude change of 0.75 mags!).

Perhaps with a customized PSF photometry package, the greater stability of the PSF photometry can be taken advantage of. However, for the current situation, the greater statistical error of the PSF photometry on the brightest stars in the field degrades its performance too much to be effective for determin-

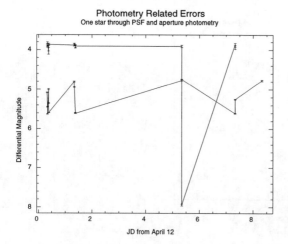

Figure 8. Light curves for a star suffering from poor photometry, for both PSF and aperture photometry.

ing planetary occultations of a program star. However, it is much better for the dimmer stars or for stars with nearby neighbors.

The two remaining stars detected as variables were thought to possibly be variable stars. They correspond to stars 33 and 58 of R.L. Gilliland, *et. al.* (1991) and are two of the three W Ursa Majoris systems discussed in that paper. Their light curves from this data are shown in figures 9 and 10. The third system is out of the field of view.

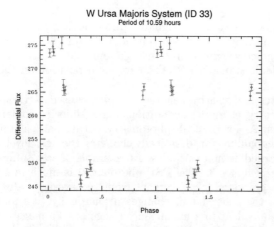

Figure 9. Light curve for the W Ursa Majoris system ID 33 (Eggen-Sandage III-2) with a period of 10.59 hours and a V amplitude of 0.05 magnitudes.

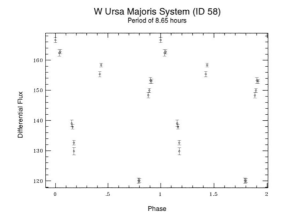

Figure 10. Light curve for the W Ursa Majoris system ID 33 (Eggen-Sandage III-33) with a period of 8.65 hours and a V amplitude of 0.18 magnitudes.

The light curves of several non-variable stars are presented in figure 11. There appear to be a few faint systematic trends (*eg* changes from night to night) but the data set is also very limited for such an analysis. The sensitivity of this technique shows here; there is only one event beyond three sigma (and it is for a low s/n star), suggesting that if the photometry can be massaged to stop the errors discussed above, an accuracy of 0.3% will provide an on-the-fly detection of a 1% transit event with a single exposure.

6. Conclusions

We recommend the following steps for any wishing to do accurate differential photometry:

(1) Carefully choose your targets; ideal are those for which the photometry method of choice, like aperture photometry, will have few problems, while still having enough stars of similar magnitude to the program star and through a range of x, y, and color indices to allow for empirical corrections for second order extinction and color effects. If such a field can't be found, be prepared to make many standard star measurements throughout the night to determine the coefficients for the extinction.

(2) Spend enough time understand the detector to calibrate out any structure (*eg.* that due to bias structure, overscan variation) and to provide as uniform a response as possible. The latter requires accurate flat fielding as well as understanding any possible corrections due to the shutter or deferred charge. As the flat fields change from night to night, and, for that matter, with a filter change, take flat fields at least once a night.

(3) Investigate the photometry you will be performing. What is the best aperture size? How is the sky to be determined? How will radiation events be handled? How are nearby neighbors accounted for?

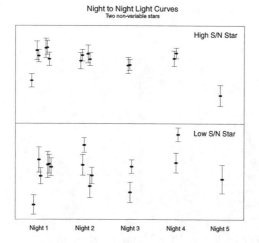

Figure 11. Night to night light curves for two non-variable stars, one high signal to noise and one low signal to noise. The error bars are photometry error.

(4) Correct for the extinction effects, via an empirical method or an analytic method. Choose a good set of comparison stars, and don't be satisfied with the first set!

Certain issues, especially the light contamination due to nearby stars and good sky background subtraction, remain to be investigated. For the former, we suggest a possible hybrid routine, in which a PSF is determined and subtracted from neighboring stars, but then aperture photometry is applied to the target star. This allows for a less accurate PSF while maintaining the high precision and robustness of aperture photometry. As for good sky subtraction, another option is to do PSF subtractions of all stars on the field, and then do a statistical averaging over subsections of the field, iteratively removing bad points (*eg.* those due to the residuals of stars). Each subsection could then be used to fit a cubic spline, and the cubic spline used for the final background determination for the field, similar in some respects to the method of Gilliland and Brown (1988).

And, for the GNAT 0.5m prototype telescope in particular, several addition considerations must be examined. In particular, the gain of the detector should be changed to take advantage of the full dynamic range of the CCD, and to minimize quantization noise. The shutter speed and possible effects due to different exposure times across the CCD should be investigated, as well as deferred charge corrections. Finally, a larger CCD (say, 2048^2) over the same field, at a site with a factor of two better seeing would yield a FWHM of approximately 5 pixels, ideal for both good aperture and PSF photometry.

Nonetheless, it is possible, using the current GNAT 0.5m prototype telescope, to achieve a night to night accuracy of 0.5% or better for a significant number of program stars and aperture photometry. If an appropriate field can be chosen (enough comparison stars of good signal to noise, without having a crowded field), it should be possible to begin a planetary transit search program with the already existing system.

References

M.S. Giampapa, E.R. Craine, D.A. Hott (1995) Icarus 118, 199-210

A.T. Young, *et. al.* (1991) PASP 103, 221

A.T. Young (1974) in *Methods of Experimental Physics,* Vol. 12A, ed. N. Carleton

R.L. Gilliland and T.M. Brown (1988) PASP 100,754

R.L. Gilliland, *et. al.* (1991) AJ 101(2),541

K.A. Montgomery, L.A. Marschall, and K.A. Janes (1993) AJ 106(1),181

CCD Strømgren and H_β Photometry of Open Clusters

N. Kaltcheva

School of Physics and Astronomy, University of St Andrews, North Haugh, St Andrews, Fife, KY16 9SS, Scotland, UK

M. I. Andersen

Nordic Optical Telescope Apartado 474, Santa Cruz de La Palma, E-38700 SPAIN

H. Jønch-Sørensen and A. N. Sørensen

Niels Bohr Institute for Astronomy, Physics and Geophysics, Juliane Maries Vej 30, 2100 Copenhagen 0, Denmark

Abstract. This is a preliminary report on $uvbyH_\beta$ photometry of the open cluster Trumpler 16, part of an extensive program of Strømgren and H_β photometry of young open clusters in the Milky Way.

1. Introduction

A project of accurate Strømgren and H_β photometry of young open clusters in the Milky Way is undertaken with general aim of providing a more detailed study of the structure of selected star forming regions. It is a common opinion that good estimates of the basic parameters do not exist for most of the young stellar groups in the Milky Way. In some cases the reason is photometry with insufficient accuracy, as noted by Crawford (1994a, 1994b). In other cases this is due to the analysis of the photometric data in terms of stellar parameters.

In this sense we are aiming at precise and homogeneous photometric data and their reliable analysis. We expect to obtain more accurate estimates of the cluster parameters and, in particular, their distances because the $uvbyH_\beta$ photometry provides useful information about the stellar physical properties in a highly efficient way. The results of the recent Hipparcos mission allow the distance calibrations in the $uvbyH_\beta$ system to be tested. On the one hand, the Hipparcos parallax data confirmed the reliability of the photometric $uvbyH_\beta$ distances for B III-IV type stars (Kaltcheva & Knude, 1998). On the other hand, they stressed the uncertainties of the spectroscopic distances (Gómez et al. 1997, Jaschek & Gómez 1997, Kaltcheva 1998), which often has been used to determine the basic parameters of many young open clusters.

Our goal is also to extend the number of young open clusters for which more complete photometric $uvbyH_\beta$ data are available. Usually such clusters have been studied in the $uvbyH_\beta$ system photoelectrically, which restricts the sample to the brightest cluster members. Our recent studies on similar objects show that the $uvbyH_\beta$ photometry and the calibrations in this system give a good agreement between the CM diagram, HR diagram, and the distance estimates and the approach is suitable for studying the structure of the star forming regions. In some of these regions a projected overlap of the young stellar aggregates often exists and a significant number of young background and foreground stars can be found within the same field. In these cases the spatial separation of the different stellar groups according to that obtained from the $uvbyH_\beta$ photometry distances is more obvious than their distinction on the photometric diagrams. An approach that deals with the individual stars, and for each star to obtain the reddening, distance, and age, is often successful in the derivation of the characteristics of the apparent aggregates.

2. Observations

About 20 young clusters in the regions of Monoceros, Vela, and Carina are included in the program so far. The brightest members of the majority of them, which are not suitable for CCD observation, are already measured in the $uvbyH_\beta$ system photo-electrically. Our project is aimed at extending the photometry to the fainter stars and to provide more complete knowledge of the properties of the young aggregates using all data. Many of the clusters in our program contain several very bright stars in the field or are located in nebulosities. This adds to the noise of the CCD data and complicates the reduction procedure. Our present effort is mainly aimed at estimating the accuracy of the CCD photometry which can be obtained for these types of objects, rather than presenting the final photometry. Here we present some of the preliminary results on Trumpler 16, one of the clusters included in the program.

The observations were carried out using the wide field instrument Danish Faint Object Spectrograph & Camera (DFOSC; see Andersen et al, 1995) on the Danish 1.54m telescope, located at the ESO - La Silla observatory in Chile. DFOSC is a combined focal reducer and low dispersion spectrograph, equipped with a thinned 2048 by 2048 pixel Loral CCD. Pixel scale is 0.39 arcsec/pix, yielding a total field of view of 13.3 x 13.3 arcmin. However the effective unvignetted field of view was limited to a central area of approximately 10 arcmin in diameter. The filter wheel inside DFOSC is slightly tilted in order to minimize the effect of 'sky-concentration' caused by internal reflections in the camera. This implies that filters of inter-mediate to narrow band width can not be mounted inside DFOSC because the tilt means that the peak transmission wavelength varies across the field. The filters must be mounted in the Filter And Shutter Unit (FASU) and since only 60mm diameter filters were available at the time of observation this resulted in a vignetting of the field of view.

Exposures were taken in the u, v, b, y, $H_{\beta_{narrow}}$ and $H_{\beta_{wide}}$ filters. Exposure times were short (typically less than 60 sec in y and 300 sec in $H_{\beta_{narrow}}$) to avoid saturation of all but the very brightest stars. At least 6 frames in each filter were obtained. Most of the primary standard stars for $uvbyH_\beta$ are too bright to be observed with a 1.54m telescope, because they saturate the CCD using even very short exposure times. For a programme like the one presented here, it is necessary to use fainter stars (V 8-15) with published photometry that is calibrated as tightly to the standard system as possible. Few such sources exist, but we have drawn our sample of secondary standard stars from Jønch-Sørensen

(1993) and Kilkenny & Laing (1992). These two sources supply photometry for a sufficient number of stars with a reasonable spread in spectral type and luminosity class and the photometry is in nice agreement (see Jønch-Sørensen, 1993) and closely transformed to the standard $uvbyH_\beta$ system as defined by the primary standard stars.

3. Data Reduction

Reductions were performed in the IRAF environment. The raw CCD frames were processed by zero-level subtraction and flat field division. The flats were short exposures of the twilight or dawn sky and super flats in each filter were created. Bad columns were masked using the IMREPLACE routine. Bad pixels were filled with the value of the median, computed in a box of 10x10 pixels. After this procedure the sky in the object frames was flat to better than 0.8 %.

Photometry was extracted using the DAOPHOT/ALLSTAR package (Steton, 1987, 1995). Stars were identified in the deepest frame with the FIND subroutine, then all other frames were aligned to it and the aperture photometry was done using the PHOT subroutine. A point spread function (PSF) linearly varying with the position on the frame was used. The PSF was modeled with a Moffat function. The parameters of the PSF were derived for each frame. About 10 bright isolated stars were chosen for the construction of the PSF. ALLSTAR was then applied, followed by subtracting all the fitted stars with the exception of the PSF stars and a new PSF was determined. Then a list of stars on a common numbering system was produced in each frame.

In order to estimate the errors in the photometry, a comparison was made between all frames in every filter. We found that all magnitudes are self-consistent over the whole range. The internal errors in the magnitudes were calculated by finding the average difference between the magnitudes, measured on different frames, after converting them to the deepest frame. These errors are of the same order in the Strømgren filters (0.009 mag in u, 0.007 mag in v, 0.007 mag in b, 0.006 mag in y), and larger in the H_β filters (0.014 mag in $H_{\beta_{narrow}}$ and 0.011 mag in $H_{\beta_{wide}}$). The internal accuracy of the photometry is estimated to be about 1 % in $(b-y)$, c_1 and m_1 and about 2% in H_β. An accuracy of this order in H_β does not allow the use of the calibrations in terms of stellar parameters in a reliable way. Apparently more H_β frames should be involved in the reduction.

At that point colour equations and zero point corrections were applied with the help of a set of stars with $uvbyH_\beta$ electro-photometry in the field. For our present purpose this calibration is sufficient but, of course, this is not a precise transformation to the standard system.

4. Preliminary Results

In Figure 1 the de-reddened colour-magnitude diagram for the field of Trumpler 16 is presented. To build this diagram, about 100 stars with electro-photometry (marked with open symbols) and 130 stars with CCD photometry (marked with filled symbols) are used. The correction of the reddening is performed with the calibration of Crawford (1978) for the B III-V type stars and with that of

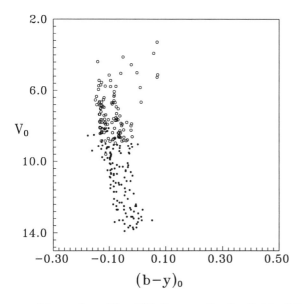

Figure 1. The CM diagram for the field of Trumpler 16.

Kilkenny & Whittet (1985) for the Supergiants. The bright stars in the field of Trumpler 16 have been observed photoelectrically in the $uvbyH_\beta$ by Kaltcheva & Georgiev (1993) and Kaltcheva (1998) and the structure of this region has been studied. We find a good agreement between the results of these previous studies and the new data.

Acknowledgments. Royal Society/NATO grant is acknowledged.

References

Andersen, J., Andersen, M. I., Klougart, J., et al., 1995, ESO Messenger 79, 12
Crawford, D. L, 1978, AJ, 83, 48
Crawford, D. L, 1994a, PASP, 106, 397
Crawford, D. L, 1994b, Rev. Mex. A&A, 29, 115
Gómez, A. E., Luri, X., Mennessier, M. O., Torra, J., Figueras, F., 1997, Proceedings of the ESA Symposium 'Hipparcos - Venice '97', 13-16 May, Venice, Italy, ESA SP-402, p. 207-212
Jaschek C. and Gómez A. E., 1998, A&A330, 619
Jønch-Sørensen, H. 1993, A&AS, 102, 637
Kaltcheva, N. & Knude, J. 1998 A&A, 337
Kaltcheva, N. & Georgiev, L. 1993 MNRAS, 261, 847
Kaltcheva, N. 1998 A&A, submitted
Kaltcheva, N., 1998, A&AS, 128, 309
Kilkenny, D. & Laing, J. D. 1992, MNRAS, 255, 308

Kilkenny D. & Whittet D. C. B., 1985, MNRAS, 216, 127
Stetson, P. B. 1987, PASP, 99, 191
Stetson, P. B. 1995, DAOPHOT User's Manual

Photometric Characteristics of the Etelman Observatory in St. Thomas, US Virgin Islands

Dirk A. Aurin[1], James E. Neff, Donald M. Drost[1]

Dept. of Physics & Astronomy, College of Charleston, Charleston, SC

Abstract. The College of Charleston and the University of the Virgin Islands are engaged in a collaborative program to operate an observatory on St. Thomas, in the US Virgin Islands. The facility has been equipped for students at either institution to complete research projects. This report describes the history of this collaboration, our efforts at refurbishing the equipment and instrumentation, our preliminary site survey, our initial research project, and future plans. We are especially interested in evaluating this unique site as a candidate for a future Automated Photometric Telescope.

1. Introduction

The Etelman Observatory sits on the top of Crown Mountain, at 1556 feet (473 meters) the highest point on the island of St. Thomas, US Virgin Islands. The facility was donated to the University of the Virgin Islands (UVI) by the Etelman family. It consists of a mountain-top house with a roof-top observatory, and it lies just a short drive from the St. Thomas campus of UVI. In addition to its undeniable natural beauty (overlooking that Atlantic Ocean to the north and the Caribbean Sea to the south), we believe it has tremendous potential as an astronomical site.

Since late 1994, UVI has had a collaborative agreement for student and faculty exchange with the College of Charleston (CofC; and with its graduate and research arm known as the University of Charleston, SC). UVI is a member of the South Carolina Space Grant Consortium. While on sabbatical from CofC, Drost began the refurbishment of the observatory. Harold Nations, formerly of CofC, installed a CCD camera and filter wheel and developed plans for conducting research at the site. Hurricane Marilyn caused substantial damage to the structure in 1995, but this has now been repaired. During the fall of 1996 and all of 1997, Aurin was an exchange student at UVI. Working under the supervision of Drost and Nations, he undertook the task of refurbishing the instrumentation and evaluating the astronomical characteristics of the site. Neff became involved with the project during the fall of 1997. Aurin and Neff revised their science program, and Aurin completed the observations for his senior thesis at CofC.

---

[1]also with Division of Science & Mathematics, University of the Virgin Islands, St. Thomas

2. Equipment and Instrumentation

The telescope, a 15" (38 cm) cassegrain reflector on a bronze German equatorial mount, was built in 1961. It is essentially identical to the 15" telescope at Villanova University. It is manually pointed with the assistance of JMI's "NGC-MAX" encoders and presently has no autoguiding capability. A 6" refractor, a 6" photographic rich-field telescope, and two small finder telescopes are co-mounted. The site infrastructure includes a road, power, water, telephone line, and a sleeping room. The remainder of the house could be developed for office space, meeting rooms, or labs.

The telescope is equipped with a computer-driven filter wheel, UBVRI filters, and a SBIG ST-6 CCD. A pentium-based PC is used to control the filter wheel and CCD and for "quick look" data analysis. We are developing IDL procedures for more detailed data analysis, both on-site and at CofC. Data is transferred from the site on 1.44 MB floppy or on 100 MB Zip disks.

3. Environmental Characteristics

Besides being a developed site with all the necessary infrastructure (and a beautiful view!), Etelman observatory has several noteworthy astronomical advantages. Perhaps foremost among these is its location approximately 65° west longitude and 18° north latitude. This is further south than the island of Hawaii and the Canary Islands and further east than anywhere in the continental US or Chile. It is surrounded by ocean with a steady temperature and wind direction and minimal artificial lighting. Its altitude (over 470 meters) places it above much of the local weather, although upslope winds can cause local cloud formation and rapidly changing conditions.

We monitored the weather, sky brightness, and seeing conditions at the observatory throughout 1997. We have also obtained meteorological records covering a much longer timeline. While we are still analyzing these data, several preliminary conclusions can be drawn. On the positive side, the sky is very dark except when local fogging conditions occur. City lights from Charlotte Amalie pose no problem, although a white strobe light on a nearby tower produces detectable interference. This problem could be minimized by reducing the strobe's power level at sunset, as permitted by the regulations. Perhaps the greatest astronomical strength of the site is steady and excellent seeing, typically better than an arc second, caused by the stable pattern of trade winds over the ocean. Local clouds forming from the upslope winds can, however, cause non-photometric conditions with little warning. We are attempting to quantify the fraction of photometric time available, but our impression is that photometric conditions seldom last for an entire night.

4. Preliminary Photometric Results

Throughout October and November 1997 we obtained UBVRI photometry of the single-lined spectroscopic binary system II Peg (HD 224085; K0 V, V=7.2 to 8.3; P∼6.7 days) and suitable check stars (HD 224084 and HD 224016). Whenever possible, we also used standard stars to perform extinction measurements throughout the night. These data were obtained in support of a molecular band study of starspot parameters (Neff, O'Neal, & Saar 1995; O'Neal, Saar, & Neff 1998) using the Phoenix infrared echelle spectrograph at Kitt Peak 7-13 November 1997. Photometric results will be published later along with the infrared data. For his senior thesis, Aurin developed a suite of IDL programs for the reduction and analysis of these data. Further development of these procedures is being carried out by students at the College of Charleston.

5. Plans for Future Development of the Site

Through the on-going collaboration between CofC and UVI, we intend to continue our monitoring of the astronomical properties of the site. We plan to involve UVI students in obtaining the site-survey measurements as part of a long-term project. With further support from the South Carolina Space Grant consortium, we have recently begun a new program utilizing both the 16" reflector in Charleston and the 15" telescope at Etelman observatory to determine orbits and parallaxes of candidate near-Earth objects (NEO's). Our first observations will be obtained in the fall of 1998, and we intend to continue this as a long-term cooperative project involving students at both institutions. The telescope-CCD combinations at both facilities yield a relatively small field of view, so we have obtained a large-format CCD and we are investigating other CCD's, focal reducers, autoguiders, etc. Etelman observatory has also been considered as a site for an environmental center and public-education facility. Ultimately, depending on the results of our site survey, we would like to locate a larger, research-grade instrument at the site.

Acknowledgments. This project arose from the efforts of Don Drost and his colleagues at the University of the Virgin Islands. Harold Nations was responsible for much of the original site refurbishment and scientific definition. We gratefully acknowledge support from our administrations at UVI and CofC. Further funding has been provided by NASA through the South Carolina Space Grant Consortium, by the Research Corporation, and by the National Science Foundation.

References

Neff, J.E., O'Neal, D., Saar, S.H. 1995, ApJ, 452, 879

O'Neal, D., Saar, S.H., Neff, J.E. 1998, ApJ, 501, L73

Update on Pine Mountain Observatory (PMO), A Report to GNAT - Summer 1998

Rick Kang

Public Education/Publicity Chair, Friends of PM

Abstract. I wanted to share our experiences at Pine Mountain as we've brought remote imaging on-line, so that other GNAT members could avoid some of these pitfalls as they bring their facilities on-line for remote photometry and other observing operations specifically for educational purposes.

1. Overview

Pine Mountain Observatory's educational mission is to promote science literacy by providing schools and general public with the opportunity to practice science as a series of observations, not a collection of facts. As of October, 1997, PMO became capable of furnishing remote users with live images through the Internet via a Santa Barbara Instrument Group (SBIG) ST6 Camera with a 500 MM Maksutov lens attached, piggybacked onto our 24" telescope. Although the telescope pointing is not automated yet, this is a culmination of six years of work and finally affords K-12 Students the opportunity to obtain actual scientific data. We had a steep learning curve with software compatibility issues, but our pilot system is simple and straight-forward to use.

Our largest problem now is getting teachers to realize what the system does and the potential for student use, and surprisingly, to actually subscribe for use.

Meanwhile, the customized 1K x 1K camera for the full scale system on our 32" telescope has just been completed, has been tested in Professor Bothun's lab, and will shortly be mounted to the telescope, as work continues on this telescope's pointing and tracking mechanism. Some of Greg's comments about early tests of the camera, plus camera specs, are included towards the end of this report.

2. Design Philosophy for the 24" Piggyback System

We needed to expedite our program so that we could demonstrate a remote imaging system to schools and to potential sources of funding. After our initial experience upgrading our 32" telescope, we felt that we should sidestep the remote pointing issue in order to get a remote imaging system rapidly off the ground. Since we'd need a body at PMO in any event, to monitor weather, adjust the split dome shutter, and monitor equipment safety and performance, we figured that this person could also aim the telescope system. We elected to piggyback the CCD so that the 24" telescope would be available for visual use for our extensive summer public visitors' program, handle instrumentation for other research, and to avoid the narrow FOV at Cassegrain focus.

3. Major Advantages/Disadvantages of our Piggyback System

Our system is relatively simple and inexpensive to construct and access, using mostly commercially available parts and software. Users either request specific targets that we image at our convenience and store on our server for future retrieval, or can apply for two-hour observing windows. For live observing, we use TIMBUKTU software (by Netopia, in Alameda, California) to facilitate the actual remote control, since the software works across WIN/MAC platforms and is very inexpensive (user copy $45.00). The wide field makes object location easy and offers a larger variety of targets/projects than high magnification, although at sacrifice of imaging Planets and other targets that present small angular size (users can track moons of Jupiter, Saturn).

Users need to run WIN95 or MAC System 7.5, need TCP/IP, and should have at least ISDN bandwidth or preferably T1 as we have at PMO. TIMBUKTU gives the user direct control of the desktop at PMO, thus great operating flexibility, although poses high security risk (dealt with by using the on-site operator manual admission mode). Remote user and local operator can keep in touch by "flash messages" within TIMBUKTU, and users can download images via an FTP package within TIMBUKTU, but need to convert from SBIG format or use SBIG compatible software.

4. Software Philosophy and Issues - Compatibility Problems, Firewall Problems

We needed to have graphics screen transfer for CCD control, and since most schools still use MACs but we have WIN95 at PMO, we elected to use TIMBUKTU as the remote control software as this package works cross-platform. TIMBUKTU requires WIN95 or MAC System 7.5, but CCDOPS, SBIG's native camera control software, wasn't compatible with WIN95, so we switched to Software Bisque's SKYPRO camera control software, which we already had (this is the only WIN/SBIG compatible software we could find, SBIG is supposed to issue a WIN version of CCDOPS for older cameras, shortly). Software Bisque has a new program, CCDSOFT, available which bundles proprietary telescope pointing control functions with camera control along with remote screen control. However it isn't MAC compatible and costs considerably more than TIMBUKTU. If everyone had compatible telescope control/camera control software (one of GNAT's goals) and compatible computer systems, efforts like CCDSOFT would solve the remote operation problem. Unfortunately, incompatibilities always surface. Note that each user needs to have a copy of the screen control software in any event. We'd recommend TIMBUKTU as a viable means of accessing whatever local software controls the telescope/CCD, although installation on MACs can be tricky (several system settings need jockeying).

The last software issue that surfaced with a vengeance as soon as we began to test connections, is that most schools have a Firewall or some other sort of smut screening software in place, but their settings also block TIMBUKTU.

Many school district network offices were willing and able to adjust their settings, but several were not, so be aware that this can be a problem and must be addressed prior to any imaging sessions!

5. Automated Pointing Issues

As many of our group have found, the process to automate telescope pointing is complex. Since we would probably need a live body on site at least initially to monitor weather, reset various programs, deal with balky machinery, and oversee initial operation for safety, we felt that we might as well use this body to point the telescope, thereby bypassing an area of major technical difficulty. We elected to survey K-12 teachers in Central Oregon to see if interest could be generated to form a corps of telescope operators, and sure enough, a dozen teachers quickly responded. Since this group has a wide range of knowledge/experience with astronomy and with telescopes, we commenced a major training program this Spring and Summer. Teacher involvement helps address the major problem outlined in the paragraphs below. Weather has been the major obstacle to doing actual remote imaging so far, we have many test images of the lens-cover! (Update: 6-13-98: Weather cleared, two schools from Eugene finally actually took images remotely that evening. Although there are still a few items with TIMBUKTU we need to iron out, the imaging system worked very well.)

6. Publicity Efforts and Responses

Here is where the results vastly differed from predictions. After we provided a series of articles in statewide publications for science teachers outlining our project, the project was presented at several state in-service sessions, and then advertised in flyers that went out to 250 schools, only ONE school responded immediately for images, and two more subsequently to date. We continue our classroom outreach, visiting several hundred classrooms each year where we bring a portable CCD camera, telescope, and describe the remote imaging operation in detail, and show archived demonstration images. Students seem quite captivated by the idea. We believe that the lack of response is due to great inertia in the teachers, primarily due to time constraints and teachers' ever increasing workload. There are probably also factors of intimidation and lack of understanding.

7. User Difficulties, Need for Classroom Outreach/Staff Development

Most teachers, due to lack of background in Space Science, lack of time, and being part of the conventional "fact-based", not observationally based curriculum, are unfamiliar with what telescopes and CCDs do, have little knowledge of what targets are feasible to work with in the sky, and hence don't have any feel for what types of projects to suggest for their students. Sometimes we find that students already have good ideas for projects. However, few teachers or students are familiar with the sky and its motion and the concepts of RA and DEC, and few have done actual observing, either naked-eye or with instruments. Also, some teachers expect that there is a bevy of instrumentation available including a spectroscope. Our initial request had assumed that spectrograms could be taken. Thus, apparently there is great need to do extensive training of teachers and program "marketing" in advance of making remote observing available.

8. Where do we go from here?

We recently were awarded a grant through the Hubble Space Telescope Science Institute's IDEAS project, to facilitate remote collection of astronomical data by students. We will continue our publicity, classroom outreach, and staff development programs, also including concentrated versions for teachers/classes participating directly in the IDEAS project. We anticipate several schools piloting research projects this Fall and hope that others will follow. As this initial project with the small scale instrument goes through shakedown, the full scale project is underway. Locate further information about the piggyback remote imaging program, including how to request images or how to apply for an observing window, at http://pmo-sun.uoregon.edu/ pmo/ under the Materials for Teachers and Students heading, and locate information about Professor Bothun's Electronic Universe concept at http://zebu.uoregon.edu.

9. Update on 1Kx1K CCD for 32" Telescope, per Professor Bothun

Wednesday, May 27th: Pine Mountain CCD camera: I'm here at the UW. Camera is great and nice compact size - won't have any trouble putting it at prime focus on existing spider. Will have it installed in my lab next week. All in all, its pretty nice and if we can get the telescope to work we will be doing research with it.
    Friday, May 29th: Camera is set up in my lab and working fine does a readout in about 5 seconds (200 kpix/sec) at my current settings. I put one of the rack and pinion focussers on it and got an image, it works good.
    Thursday, June 4th: COWCAM is the following: 1024x1024 thinned, back illuminated SITe chip, 50% QE at 4000 Angstroms. Readnoise is 9 electrons when read out at 200 Kpix per second (meaning we can read the device out in 5 seconds!) Plate scale is 1.8 arcseconds per pixel. FOV = 1/2 x 1/2 degree.
    Our 32" telescope is a 1970 vintage prototype instrument designed and constructed by Sigma Research, of Pasco, Wash.. Ultimately we would like this telescope to perform in a fully automated mode, for use by students as well as for research by the professional community. We also contemplate automating our 24" Boller & Chivens at a future date. To complete the 32", we need to install an encoder on the RA mechanism that the original design didn't feature (original RA encoder was mounted on the RA drive motor!), install a new controller board for the encoders, test the revised control software, and balance the telescope after the new CCD is installed. The "sister" telescope is at Rattlesnake Ridge Observatory above Hanford, Washington, also undergoing upgrades.

National Astronomical Observatory Tonantzintla, México

J. H. Peña, R. Peniche, B. Sánchez, C. Tejada and R. Costero

Instituto de Astronomía, UNAM

1. Historical Background

In a most colorful and impressive manifestation of a nation's urge to foster purely scientific research, high officials of the Mexican government and a large delegation of American scientists, mostly astronomers, formally dedicated the new Mexican National Astrophysical Observatory near the small Aztec village of Tonantzintla on February 17, 1942.

The region selected for the new observatory is rich in cultural traditions, for it was in Cholula that the Toltecs, and later the Aztecs, centered many of their temples of learning and religion.

From the astronomical point of view, the site of the new observatory is likewise very advantageous. With a latitude of nearly 19° N, the observatory is nicely placed for observations of the whole Milky Way, and with an elevation of about 7,000 feet, it is well above the usual strata of dust and fog.

N. U. Mayall
México Dedicates a New Observatory
PASP 54, 117, 1942.

The observatory of Tonantzintla is well worth a visit by any interested amateur astronomer from the United States who happens to be down Mexico-way. For scenery, there is no place in México from which Popo and Ixta can be seen any better. The region of Tonantzintla and Cholula is one of the richest in archeological gems and beautiful churches, one of the finest being within walking distance of the observatory in Tonantzintla.

Bart J. Bok
Tonantzintla Revisited
Sky and Telescope No. 42

2. Actual Situation

The observing conditions are not as promising as they were when the observatory was opened.

The neighboring villages of Puebla and Cholula have grown fast. Puebla is up to five million inhabitants and, hence, the light pollution of the sky is severe.

The Popocatepetl is an active volcano which is going through a period of activity. Large clouds of dust and ash make astronomical observations impossible at times.

Despite all these inconveniences its closeness to México City, its observational and accommodation facilities, make it desirable and possible to continue research and educational projects that make good use of the 1m telescope.

The site

latitude	longitude	altitude
19°01'58"	98°18'50"	2147 m

climate clear in autumn and winter and cloudy in summer

The 1 m telescope

- primary mirror of 1m
- secondary mirror with f/15
- plate scale of 13.53"/mm
- guiding console:
  - positioning limits
  - hour angle between −5.6 and 5.6 hours
  - declination between −60 and 80 degrees
  - zenith angle between 0 and 75.5 degrees
- maximum velocity of guiding 1 degree/s
- guiding accuracy of the controller 1/3"
- uncertainty in the telescope positioning 1' for $Z < 60°$
- the bright star catalog is available, as well as a diskette with a preselected sequence of objects.

The ccd-mil

- the system, Photometrics, consists of control card of the camera. It is installed in a PC 486
- electronic and thermic controllers of the camera, are installed in the telescope plate
- cryostat
- users interface, in Windows environment: PMIS
- the detector is Thomson 1024 × 1024, thinned and with Metachrome II coating. The pixels are of 19 microns, with a 14 bit resolution and a reading rate of 50 KHz.
- The direct camera has a field of 4.2' by side.

UBVRI filters (Schott)			
Band	lambda (A)	width (A)	thickness (mm)
U	3582	500	4
B	4300	1000	4
V	5400	900	4
R	6400	1300	4
I	8900	3400	3

B & CH spectrograph

- gratings of 400, 600, and 830 grooves/mm
- comparison lamp of He-Ar
- camera adapted to the CCD-Mil gives a resolution of 0.7 A/pixel with the 600 grating

Scientific aims

- morphology of HII regions, Costero (costero@astroscu.unam.mx)
- remotization of the 1m telescope, Sánchez (beatriz@astroscu.unam.mx)
- variable stars in clusters, Peña & Peniche (jhpena@astroscu.unam.mx and rpeniche@astroscu.unam.mx)

3. Morphology of HII Regions

The catalogs of HII regions that have been elaborated until now have used photographic plates. These have the advantage of having wide fields, but they also have the disadvantage of their range, saturation, and wide filters (V, R, ...), which often camouflage or make the detection of emission lines sources impossible. There is only one catalog which uses narrow filters, but again it is elaborated with photographic plates and the scale of the plates does not permit detailed study.

For this reason, the construction of a narrow filter catalog allows the identification of new objects hidden on photographic plates. Objects such as H-H (following the tradition of Tonantzintla), jets, collisions of all kinds including the remains of hidden supernovas in HII regions and shells created by stellar wind can now be identified. The use of CCDs greatly facilitates the detection of these sources through image processing methods.

In order to do this kind of narrow-band filter imaging of extended emission objects, we have designed a prime-focus reducing camera for the 1-m Tonantzintla telescope. We wanted a low cost, wide field instrument, for the 500 to 960 nm wavelength range, to be used with our Thomson 1024×1024 CCD, 19×19 μm pixels. The final design yielded a four lens F/3.3 camera, with images inside a 2 × 2 pixel square on a 20' × 20' field and for the whole wavelength range.

For guiding we will use a piggy-back Dall Kirkham telescope 250 mm aperture and 3810 mm focal length, with a three-axis mounting at the focal plane for a small guiding CCD. The latter is a 2.64 mm square chip, 192 × 163 pixels, 17 × 13 μm in size, that we already own.

This camera will be constructed in our optical and mechanical shops in the near future. It will cost less than $ 1500 including a motorized filter wheel for the camera and the guiding telescope.

The possibility of students using the telescope at Tonantzintla to do practices and therefore learn observation techniques at a low cost has often been suggested. This would allow a great savings from having to send them to San Pedro Martir to observe.

4. Remotization of the 1m Telescope

In order to increase the potential of the 1m equatorial mount telescope of the Observatorio Astronómico Nacional, located at Tonantzintla for research, teaching and outreach programs, the Institute of Astronomy has generated a project for the remote operation of the telescope and the acquisition of astronomical data from the University site in México City. The telescope has a computerized control and its programs were recently optimized for remote control handling. The dome was optoelectronic codified in order to have its movements coordinated with the telescope. The Ethernet type fiber optics network is the communication channel for the remote control of the telescope. This will allow astronomers to carry out a significant number of projects. (For a detailed description see A. Bernal, B. Sánchez and A. Iriarte, 1998).

5. Variable Stars in Clusters

The advantages of studying variable stars in a given cluster are mainly that the member stars are at the same distance and that the parameters that fix the evolution of the stars, such as age and chemical composition can be considered to be the same for all the stars within the cluster. These parameters, although generally poorly determined or known indirectly for only a few stars, are well known for nearby clusters and, jointly with the mass and the effective temperature, allow a better determination of the description of the pulsation mechanisms. This is the main reason that makes the study of variables in clusters important since the differences shown will, in principle, throw light on the causes that provoke the triggering mechanism of the pulsations.

6. Methodology

Strömgren uvby-β photometry provides unique opportunities for determining both membership to the cluster and physical characteristics of the observed stars. In the desired study, besides observing the short period variable stars in differential photometry, several bright stars in the direction of the cluster as well as several standard stars are observed in order to transform their photometric values into the standard system.

This has been done at the San Pedro Mártir Observatory in the past. Table 1 presents a summary of the obtained results.

Table 1. Observed clusters in Strömgren photometry.

cluster	variables found or observed	references
NGC 2539 & 6494		Rev. Mex. Astron. Astrofis. 14, 420
NGC 6882/5	2	IBVS 3488 & ESO Proceedings 36
NGC 7062		Rev. Mex. Astron. y Astrofis. 20, 127
Coma	FM Coma	A & A 268, 123.
Coma	HR 4684 and HD 106103	Rev. Mex. Astron. Astrofis. 25, 129-141
NGC 1342		Rev. Mex. Astron. Astrofis. 28, 7-16
NGC 1444, 1662, 2129, 2169 & 7209		Rev. Mex. Astron. Astrofis. 28, 139-152
NGC 2422		A&AS 123, 25-30
Praesepe	KW 45, 204, 207, 445, 323,	A&AS 129, 9-22
NGC 2264	W2 and W20	in preparation

Cluster membership is established with the advantages of the Strömgren photometry of the cluster, a calibration by Nissen (1988) which follows previous calibrations by Crawford (1975, 1979) for the A and F stars and of Shobbrook (1984) for early type stars which have been already employed in previous analyses of open clusters (Peña and Peniche, 1994). From the distance to the stars evaluated, a mean distance and standard deviation is calculated; the criterion for membership is established as the distance within one sigma of the mean. Once this is done, average parameters such as reddening $E(b-y)$ distance and chemical composition, are determined.

Once the reddening has been determined it is possible to calculate the unreddened colors $(b-y)_0$, m_0 and c_0. Then, the location of each star is fixed at the $(b-y)_0$ vs c_0 diagram of Relyea and Kurucz (1978); from it, the surface temperatures and gravities, $\log Te$ and $\log g$, are determined for each star. Another way in which this latter quantity can be determined is through the calibrations of Petersen and Jorgensen (1972) or by the calibrations of Pérez et al. (1989).

The pulsation mode is determined from the well-known relation (Petersen and Jorgensen, 1972; Breger, 1990).

The age of the cluster is fixed after establishing physical characteristics such as $\log Te$ and $\log g$ for each star in the theoretical grids of Relyea and Kurucz (1978).

Acknowledgments. We would like to thank the funds provided by Conacyt through grant 3925E. Proofreading was done by J. Miller.

References

Bernal, A. Sánchez B. and Iriarte, A. 1998 SPIES International Symposium on Astronomical Telescopes and Instrumentation. Kona, Hawaii, USA. Proceedings of SPIE Vol 3351.

Breger M. 1990, A & A 231, 56

Crawford D. L. 1975, AJ 80, 955

Crawford D. L. 1979, AJ 84, 1858

Nissen P. 1988, A & A 199, 146

Paparo M., Peña J. H., Peniche R., Ibanoglu C. Tunca Z. and Evren S. 1993. Astron. Astrophys. 268, 123.

Peniche R., Peña, J. H. Díaz, S. H. and Gómez, T. 1990 Rev. Mex. Astron. y Astrofis. 20, 127.

Peniche R., Peña, J. H., and Garrido R. 1990 ESO Proceedings No 36.

Peña, J. H., Peniche, R., Díaz-Martínez, S.H. 1990. IBVS 3488.

Peña J. H., Peniche R., Mujica R., Ibanoglu C., Ertan A. Y., Tumer O., Tunca Z. and Evren S. 1993. Rev. Mex. Astron. Astrofis. 25, 129-141.

Peña J. H. and Peniche R.. 1994. Rev. Mex. Astron. Astrofis. 28, 139-152

Peña J. H. Peniche R. Bravo, H. and Yam O.. 1994. Rev. Mex. Astron. Astrofis. 28, 7-16.

Peña J.H., Peniche R. 1994, RevMexAA 28,139

Peña J. H. Peniche R., Hobart A., Rolland A., López de Coca P., Paparo M., Parrao L., De la Cruz C.

Olivares J. I., Costa V., Ibanoglu C. , Ertan A. Y. , Tumer O., Tunca Z. and Evren S., 1998 AAS 129, 9-22

Pérez M. R., Joner M. D., Thé P. S., Westerlund B. E., 1989 PASP 101, 195

Petersen J. O., Jorgensen H. E. 1972, A & A 17, 367

Rojo E., Peña J. H. and González, D. 1997 A & ASS 123, 25-30

Relyea L. J., Kurucz R. L. 1978, ApJs 37,45

Shobbrook R. R. 1984, MNRAS 211, 659

Author Index

Andersen, M. I. 252
Anderson, J. M. **125**
Angione, R. J. **16**
Anthony-Twarog, B. J. **192**
Aurin, D. A. **257**

Beers, T. C. 192
Bucciarelli, B. 154

Casalegno, R. **154**
Costero, R. 264
Couch, J. 111
Craine, E. R. **83**, **213**, 238
Craine, P. R. 83, **133**
Crawford, D. L. **3**, **6**, **9**, 83

Davis, D. R. 170
Davis, L. E. **35**
Drost, D. M. 257

Esquerdo, G. 170
Everett, M. E. 170

Foster, T. **111**

Garcia, J. 154
Giampapa, M. S. 238
Greimel, R. 198

Hawley, S. L. 192
Heasley, J. N. **56**
Howell, S. B. **170**
Hube, D. 111

Jønch-Sørensen, H. 252

Kaltcheva, N. **252**
Kang, R. **260**

Langill, P. 103
Lasker, B. M. 154

Martin, B. 111
Martinez, P. 213
Mattox, J. R. **95**
Mighell, K. J. **50**
Milone, E. F. **103**

Neff, J. E. 257
Newberry, M. V. **74**

Ostrowski, T. A. **207**

Peña, J. H. **264**
Peniche, R. 264

Robb, R. M. **198**
Routledge, D. 111

Sánchez, B. 264
Sarajedini, A. 192
Snowden, M. 213
Sørensen, A. N. 252
Stencel, R. E. 207

Taylor, J. M. **238**
Tejada, C. 264
Tucker, R. A. **24**
Twarog, B. A. 192

Van Lew, T. 170
Vaneldik, F. 111

Weidenschilling, S. 170
Wray, J. D. **143**

Subject Index

3C 279, 95

AIT, 83
aperture corrections, 50
aperture photometry, 35, 50, 154, 213
aperture size, 50
APPHOT, 35
ATIS, 83
atmospheric extinction, 3
atmospheric transmission, 16
automatic telescope, 83, 143
Automatic Telescope Instruction Set, 83

bias frames, 35
BL Cam, 111
BL Lac, 95
BL Lac objects, 95
blazars, 95
BY Dra, 198

CCD camera, 24, 83, 111, 133, 170, 207, 257
CCD cooling, 24, 133
CCD device, 24, 111
CCD photometry, 35, 83
CCD quantum efficiency, 24
CCD reduction, 35, 154, 252
CCDPHOT, 35, 192
CCDRED, 35
CCDTIME, 35
cluster photometry, 56, 103, 264
color-magnitude diagram, 56
cosmic ray removal, 35, 154
crowded fields, 35, 56
cryogenic coating, 24
cryopumping, 24
cryostat, 24, 133
CTB-87, 111

DAOFIND, 35
DAOPHOT, 35, 56
dark subtraction, 35
DC offsets, 35
differential photometry, 238

El Chichon, 16

extinction, 16
extra-solar planets, 170

filter wheel radiometry, 16
FITS, 35
flat-fielding, 35, 125, 154, 213

gain, 74
GNAT, 9, 16, 83, 95, 133, 143, 213
GNATCS, 83
growth curve photometry, 213
GSC 3492_742, 198
GSC 3493_1097, 198
GSC 3493_1158, 198

H II region, 264
high gain values, 74
Hipparcos, 207
HST, 56

ICE, 35
II Peg, 257
image headers, 35
instrumental magnitudes, 50
interstellar absorption, 3
IRAF, 35, 83, 95, 154, 170, 198, 238

Joule-Thompson refrigerator, 24

LOWTRAN, 16
lunar photometry, 125

M51, 103
M57, 111
M67, 213, 238
M92, 56, 103
MDHS, 35
mechanical cooling, 24
Mie scattering, 16
MIRA, 74, 83, 213
Mira stars, 83, 207
MK spectral classification, 9
molecular absorption, 16
mosaicked images, 35
Mount Laguna Observatory, 16
multi-pinned-phase, 24, 125

NGC6637, 56
noise, 74, 133

observing planning, 35
open cluster photometry, 252
open clusters, 6
outgassing, 24

photometric accuracy, 6, 74
photometric error, 6, 74, 213
photometric errors, 35
photometric precision, 3, 6, 16, 74, 238
photometric reductions, 213
photometric systems, 9
Pinatubo, 16
planetary nebulae, 103
planetary transits, 170
point spread function, 35, 50, 56
PSF, 35, 50, 56, 154, 170, 238
PSF photometry, 95
PSF spatial variation, 56

quantization losses, 74
quantization noise, 74
quasars, 95

R Leo, 207
radio frequency interference, 125
Rayleigh scattering, 16
remote observing, 260, 264

SAOIMAGE, 35
scintillation, 16, 170
SDS, 83
signal-to-noise, 50
sky background, 3, 35, 50, 56, 74, 125
solar constant, 16
spectroscopy, 9
SPS, 56
standard stars, 9, 35
star centering, 50, 56
star finding, 35, 56
stellar photometry, 35, 83, 192, 207, 257
stray light, 111
Stromgren system, 192
STSDAS, 35
STSDAS SYNPHOT, 35
super insulation, 24
SX Phe, 111

T Hya, 207

thermal dark current, 24
thermoelectric cooling, 24, 133
transformation equations, 35

vacuum, 24, 133

W Uma, 198
WEB telescope, 95

X Hya, 207
XIMTOOL, 35